OpenIntro: Statistics
Preliminary Edition

David M Diez
Postdoctoral Fellow
Department of Biostatistics
Harvard School of Public Health
david.m.diez@gmail.com

Christopher D Barr
Assistant Professor
Department of Biostatistics
Harvard School of Public Health
cdbarr@gmail.com

Mine Çetinkaya
Graduating Doctoral Student
Department of Statistics
University of California, Los Angeles
cetinkaya.mine@gmail.com

ISBN: 1-45370-097-8

Contents

Preface

OpenIntro is a small organization focused on developing free and affordable education materials. This preliminary textbook, *OpenIntro: Statistics*, is a pilot project and is intended for an introductory college course in statistics for non-majors. A PDF version of the book is available online at `openintro.org` at no cost, and the book's source will be released in 2011 under a ShareAlike license. The chapters are as follows:

1. **Introduction to data.** Introduce data structures, variables, summaries, graphics, and basic data collection techniques.

2. **Probability.** This chapter discusses the basic principles of probability. An understanding of this chapter is not required for Chapters 3-7.

3. **Distributions of random variables.** The normal model and other key distributions are introduced here.

4. **Foundations for inference.** This chapter is devoted to laying out the general ideas for inference in the context of means in the one sample case.

5. **Large sample inference.** One and two sample inference for means and proportions when sufficiently large samples are utilized to allow for the normal model.

6. **Small sample inference.** Inference using the t distribution for means under the one and two sample cases.

7. **Introduction to regression.** A introduction to the ideas of regression principles. This chapter could be introduced anytime after Chapter 1.

Sections marked in the Table of Contents with an asterisk (*) are optional and may rely on Chapter 2.

The authors would like to thank Filipp Brunshteyn, Meenal Patel, and Rob Gould for their time and contributions. Without their help, this book would not be possible.

To keep up-to-date on the project or to provide feedback, please see

`openintro.org`

Chapter 1

Introduction to data

Scientists seek to answer questions using rigorous methods and careful observations. These observations – collected from the likes of field notes, surveys, and experiments – form the backbone of a statistical investigation and are called **data**. Statistics is the study of how best to collect, analyze, and draw conclusions from data. It is helpful to put statistics in the context of a general process of investigation:

1. Identify a question or problem.
2. Collect relevant data on the topic.
3. Analyze the data.
4. Form a conclusion.

Statistics as a subject focuses on making stages (2)-(4) objective, rigorous, and efficient. That is, statistics has three primary components: How best can we collect the data? How should it be analyzed? And what can we infer from the analysis?

The topics scientists investigate are as diverse as the questions they ask. However, many of these investigations can be addressed with a small number of data collection techniques, analytic tools, and fundamental concepts in statistical inference. This chapter provides a glimpse into these and other themes we will encounter throughout the rest of the book. We introduce the basic principles of each branch and learn some tools along the way. We will encounter applications from other fields, some of which are not typically associated with science but nonetheless can benefit from statistical study.

1.1 An example: treating heart attack patients with a new drug

In this first section, we consider an experiment used to evaluate whether a drug, sulphinpyrazone, reduces the risk of death in heart attack patients. In this study, we might start by writing the principle question we hope to answer:

> Does administering sulphinpyrazone to a heart attack patient reduce the risk of death?

The researchers who asked this question collected data on 1475 heart attack patients. Each volunteer patient was randomly assigned to one of two groups:

> **Treatment group**. Patients in the treatment group received the experimental drug, sulphinpyrazone.
>
> **Control group**. Patients in the control group did not receive the drug but instead were given a **placebo**, which is a fake treatment that is made to look real.

In the end, there were 733 patients in the treatment group and 742 patients in the control group. The patients were not told which group they were in, and the reason for this secrecy is that patients who know they are being treated often times show improvement (or slower degeneration) regardless of whether the treatment actually works. This improvement is called a **placebo effect**. If patients in the control group were not given a placebo, we would be unable to sort out whether any observed improvement was due to the placebo effect or the treatment's effectiveness.

After 24 months in the study, each patient was either still alive or had died; this information describes the patient's **outcome**. So far, there are two characteristics about each patient of relevance to the study: patient `group` and patient `outcome`. We could organize this data into a table. One common organization method is shown in Table 1.1, where each patient is represented by a row, and the columns relate to the information known about the patients.

Considering data from each patient individually would be a long, cumbersome path towards answering the original research question. Instead, it is often more useful to perform a data analysis, considering all of the data at once. We first

Patient	`group`	`outcome`
1	`treatment`	`lived`
2	`treatment`	`lived`
⋮	⋮	⋮
1475	`control`	`lived`

Table 1.1: Three patients from the sulphinpyrazone study.

		outcome		
		lived	died	Total
group	treatment	692	41	733
	control	682	60	742
	Total	1374	101	1475

Table 1.2: Descriptive statistics for the sulphinpyrazone study.

might summarize the raw data in a more helpful way, like that shown in Table 1.2. In this table, we can quickly see what happened over the entire study. For instance, to identify the number of patients in the treatment group who died, we could look at the intersection of the `treatment` row and the `died` column: 41.

⊙ **Exercise 1.1** Of the 733 patients in the treatment group, 41 died. Using these two numbers, compute the proportion of patients who died in the treatment group. Answer in the footnote[1].

We can compute summary statistics from the summary table. A **summary statistic** is a single number summarizing a large amount of data[2]. For instance, the primary results of the study could be placed in two summary statistics: the proportion of people who died in each group.

Proportion who died in the treatment group: $41/733 = 0.056$.

Proportion who died in the control group: $60/742 = 0.081$.

These two summary statistics are useful in evaluating whether the drug worked. There is an observed difference in the outcomes: the death rate was 2.5% lower in the treatment group. We will encounter many more summary statistics throughout this first chapter.

Here we now move into the fourth stage of the investigative process: drawing a conclusion. We might ask, Does this 2.5% difference in death rates provide convincing evidence that the drug worked? Even if the drug didn't work, typically we would not get the exact same death rate in each group. Maybe the difference of 2.5% was just due to chance. Regrettably, our analysis does not indicate whether what we are seeing is real or a natural fluctuation. We will have to wait until a later section before we can make a more formal assessment.

We perform a more formal analysis in Section 1.6 (optional) for this drug study so that we can draw a more careful conclusion from the data. However, this analysis will not make much sense before we discuss additional principles, ideas, and tools of statistics in Sections 1.2-1.5.

[1]The proportion of the 733 patients who died is $41/733 = 0.056$.

[2]Formally, a summary statistic is a number computed from the data. Some summary statistics are more useful than others.

1.2 Data basics

Effective presentation and description of data is a first step in most analyses. This section introduces one structure for organizing data and also terminology that will be used throughout this book.

1.2.1 Observations, variables, and cars

Table 1.3 displays rows 1, 2, 3, and 54 of a data set concerning cars from 1993. These observations (measurements) of 54 cars will be referred to as the `cars` data set.

	type	price	mpgCity	driveTrain	passengers	weight
1	small	15.9	25	front	5	2705
2	midsize	33.9	18	front	5	3560
3	midsize	37.7	19	front	6	3405
⋮	⋮	⋮	⋮	⋮	⋮	⋮
54	midsize	26.7	20	front	5	3245

Table 1.3: The `cars` data matrix.

Each row in the table represents a single car or **case**[3] and contains six characteristics or measurements for that car. For example, car 54 is a midsize car that can hold 5 people.

Each column of Table 1.3 represents an attribute known about each case, and these attributes are called **variables**. For instance, the `mpgCity` variable holds the city miles per gallon rating of every car in the data set.

In practice, it is especially important to ask clarifying questions to ensure important aspects of the data are understood. For instance, units are not always provided with data sets, so it is important to ask. Descriptions of all six car variables are given in Table 1.4.

⊙ **Exercise 1.2** What are the units of the `price` variable? Table 1.4 will be helpful.

1.2.2 Data Matrix

The data in Table 1.3 represent a **data matrix**, which is a common way to organize data. Each row of a data matrix represents a separate case and each column represents a variable. A data matrix for the drug study introduced in Section 1.1 is shown in Table 1.1 on page 2, where patients represented the cases and there were two recorded variables.

Data matrices are convenient for recording data as well as analyzing data using a computer. In data collection, if another individual or case is added to the data

[3]A case may also be called an **observational unit**.

variable	description
type	car type (small, midsize, or large)
price	the average purchase price of the vehicles in $1000's (positive number)
mpgCity	rated city mileage in miles per gallon (positive number)
driveTrain	the drive train (front, rear, 4WD)
passengers	passenger capacity (positive whole number, taking values 4, 5, or 6)
weight	car weight in pounds (positive number)

Table 1.4: Variables and their descriptions for the cars data set.

set, an additional row can be easily added. Similarly, additional columns can be added for new variables.

⊙ **Exercise 1.3** Researchers collected body measurements for bushtail possums in Eastern Australia. They trapped 104 possums and recorded age, gender, head length, and four other pieces of information for each possum. How might this data be organized in a data matrix? Answer in the footnote[4].

1.2.3 Variable types

Examine each of the following variables in the cars data set: type, price, driveTrain, and passengers. Each of these variables is inherently different from the others yet many of them share certain characteristics.

First consider price, which is said to be a **numerical** variable since the values it takes are numbers and those numbers have a meaningful ordering. That is, all 54 cars could be ordered according to price, and this ordering would have meaning (e.g. least expensive to most expensive).

The passengers variable is also numerical, although it seems to be a little different than price. The variable passengers can only take whole positive numbers (1, 2, ...) since there isn't a way to have 4.5 passengers. The variable passengers is said to be **discrete** since it only can take numerical values with jumps (e.g. 1, 2, 3, ...). On the other hand, price is said to be **continuous**.

The variable driveTrain can only take a few different values: front, rear, and 4WD. Because the responses themselves are categories, driveTrain is called a **categorical** variable[5]. The three possible values (front, rear, 4WD) are called the variable's **levels**.

Finally, consider the type variable describing whether a car is small, medium, or large. This variable seems to be a hybrid: it is a categorical variable but the

[4]Here each possum represents a case, and there are seven pieces of information recorded for each case. A table with 104 rows and seven columns could hold this data, where each row represents a possum and each column represents a particular type of measurement or recording.

[5]Sometimes also called a **nominal** variable.

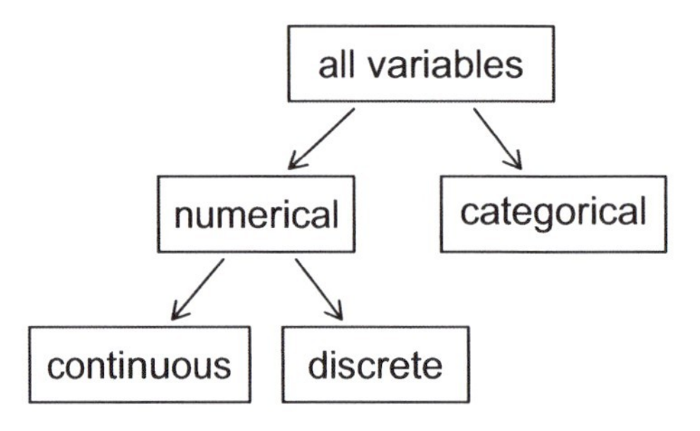

Figure 1.5: Breakdown of variables into their respective types.

levels have some inherent ordering. A variable with these properties is called an **ordinal** variable. To simplify analyses, any ordinal variables in this book will be treated as categorical variables.

● **Example 1.1** Data was collected about students in an introductory statistics course. For each student, three variables were recorded: number of siblings, student height, and whether the student had previously taken a statistics course. Classify each of the variables as continuous numerical, discrete numerical, or categorical.

The number of siblings and student height represent numerical variables. Because the number of siblings is a count, it is discrete. Height varies continuously, so it is a continuous numerical variable. The last variable classifies students into two categories – those who have and those who have not taken a statistics course – which makes this variable categorical.

⊙ **Exercise 1.4** In the sulphinpyrazone study from Section 1.1, there were two variables: `group` and `outcome`. Are these numerical or categorical variables?

1.2.4 Variable relationships

Many analyses are motivated by a researcher looking for a relationship between two or more variables. A biologist studying possums in Eastern Australia may want to know answers to some of the following questions.

(1) If a possum has a shorter-than-average head, will its skull width usually be smaller or larger than the average skull width?

(2) Will males or females, on average, be longer?

(3) Which population of possum will be larger on average: Victoria or the other locations they live?

(4) Does the proportion of males differ based on location, i.e. from Victoria to the other locations?

Figure 1.6: The common brushtail possum of Australia[a].

[a]Photo by wollombi on Flickr: `http://flickr.com/photos/wollombi/58499575/`

To answer these questions, data must be collected. Four observations from such a data set is shown in Table 1.7, and descriptions of each variable are presented in Table 1.8. Examining summary statistics could provide insights to each of the four questions about possums. Additionally, graphs of the data are useful to answering such questions.

Scatterplots are one type of graph used to study the relationship between two variables. Figure 1.9 compares the `headL` and `skullW` variables. Each point represents a single possum. For instance, the red dot corresponds to Possum 1 from Table 1.7, which has a head length of 94.1mm and a skull width of 60.4mm. The scatterplot suggests that if a possum has a short head, then its skull width also tends to be smaller than the average possum.

⊙ **Exercise 1.5** Examine the variables in the `cars` data set, which are described in Table 1.4 on page 5. Create two questions about the relationships between these variables that are of interest to you.

1.2.5 Associated and independent variables

The variables `headL` and `skullW` are said to be *associated* because the plot shows a discernible pattern. When two variables show some connection with one another, they are **associated** variables. Associated variables can also be called **dependent** variables and vice-versa.

	pop	sex	age	headL	skullW	totalL	tailL
1	Vic	m	8	94.1	60.4	89.0	36.0
2	Vic	f	6	92.5	57.6	91.5	36.5
3	Vic	f	6	94.0	60.0	95.5	39.0
⋮	⋮	⋮	⋮	⋮	⋮	⋮	⋮
104	other	f	3	93.6	59.9	89.0	40.0

Table 1.7: Four lines from the `possum` data set.

variable	**description**
`pop`	location where possum was trapped (`Vic` or `other`)
`sex`	possum's gender (`m` or `f`)
`age`	age, in years (whole number, data range: `1` to `9`)
`headL`	head length, in mm (data range: `82.5` to `103.1`)
`skullW`	skull width, in mm (data range: `50.0` to `68.6`)
`totalL`	total length, in cm (data range: `75.0` to `96.5`)
`tailL`	tail length, in cm (data range: `32.0` to `43.0`)

Table 1.8: Variables and their descriptions for the `possum` data set.

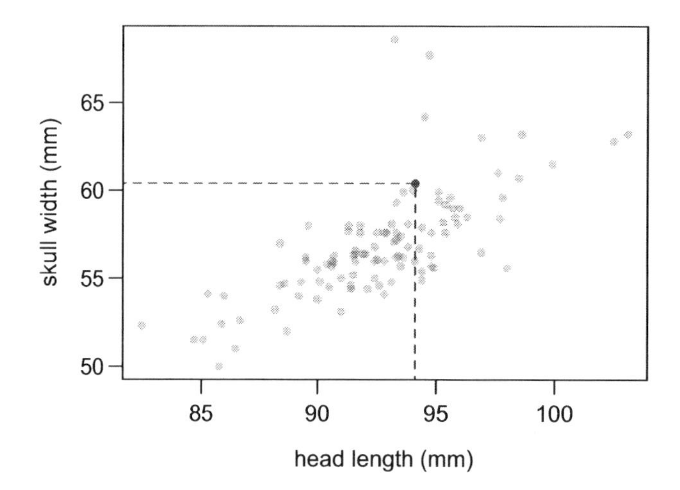

Figure 1.9: A scatterplot showing `skullW` against `headL`. The first possum with a head length of 94.1mm and a skull width of 60.4mm is highlighted.

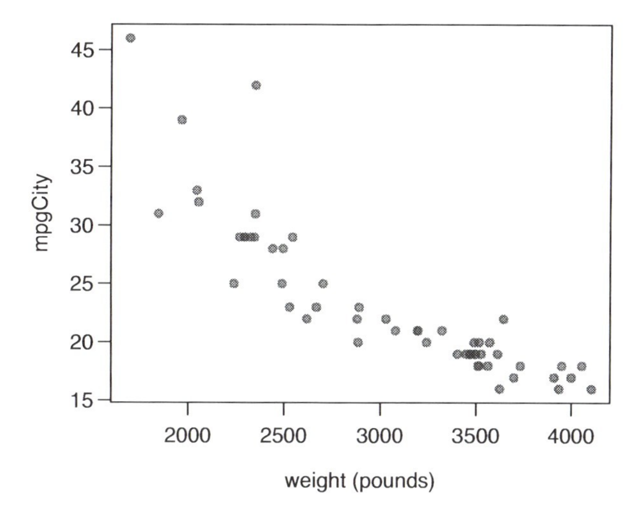

Figure 1.10: A scatterplot of `mpgCity` versus `weight` for the `cars` data set.

● **Example 1.2** Examine the scatterplot of `weight` and `mpgCity` in Figure 1.10. Are these variables associated?

It appears that the heavier a car is, the worse mileage it gets. Since there is some relationship between the variables, they are associated.

Because there is a downward trend in Figure 1.10 – larger weights are associated with lower mileage – these variables are said to be **negatively associated**. A **positive association** is shown in the possum data represented in Figure 1.9, where longer heads are associated with wider skulls.

If two variables are not associated, then they are **independent**. That is, two variables are independent if there is no evident connection between the two. It is also possible for cases – such as a pair of possums or a pair of people – to be independent. For instance, if possums 1 and 2 do not have any natural connection, such as being siblings or competing for resources in the same territory, then they can be called independent.

Associated or independent, not both
A pair of variables are either related in some way (associated) or not (independent). No pair of variables is both associated and independent. These same definitions hold true for a pair of cases as well.

Variables can be associated or independent. Cases can be associated or independent. However, a variable cannot be associated or independent of a case. For example, the `headL` variable cannot be independent of possum 1. In statistics, for two things to be independent of each other, they must be comparable. It makes no sense to discuss independence between a variable and a case.

Association between categorical variables will be discussed in Section 1.4, and associations between categorical and numerical variables will be discussed specifically in Section 1.4.5.

1.3 Examining numerical data

The `cars` data set represents a *sample* from a larger set of cases. This larger set of cases is called the **population**. Ideally data would be collected from every case in the population. However, this is rarely possible due to high costs of data collection. As a substitute, statisticians collect subsets of the data called **samples** to gain insights into the population. The `cars` data set represents a sample of all cars from 1993, and the `possum` data set represents a sample from all possums in the Australian states of Victoria, New South Wales, and Queensland. In this section we introduce summary statistics and graphics as a first step in analyzing numerical data from a sample to help us understand what is going on in the population as a whole.

1.3.1 Scatterplots for paired data

A **scatterplot** provide a case-by-case view of data for two numerical variables. In Section 1.2.4, a scatterplot was informally introduced and used to examine how head length and skull width were related in the `possum` data set. Another scatterplot is shown in Figure 1.11, comparing `price` and `weight` for the `cars` data set.

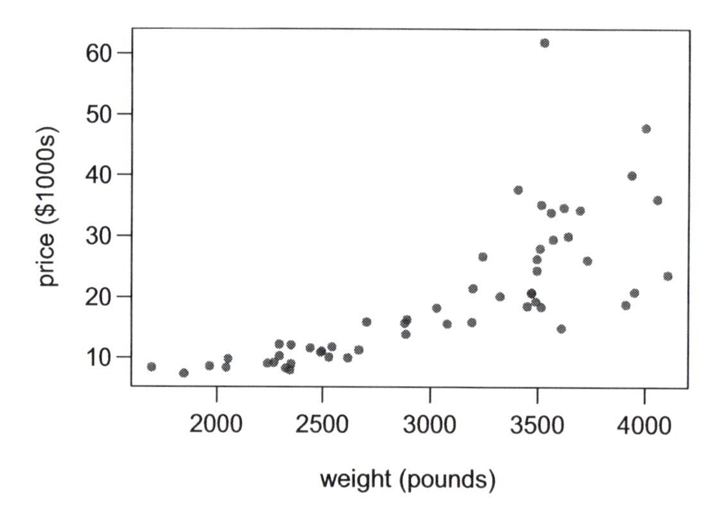

Figure 1.11: A scatterplot of `price` versus `weight` for the `cars` data set.

In any scatterplot, each point represents a single case. Since there are 54 cases in `cars`, there are 54 points in Figure 1.11.

⊙ **Exercise 1.6** What do scatterplots reveal about the data, and how might they be useful?

Some associations are more linear, like the relationship between `skullW` and `headL`, shown in Figure 1.9 on page 8. Others, like the one seen in Figure 1.10 on page 9, can be curved.

⊙ **Exercise 1.7** Describe two variables that would have a horseshoe shaped association in a scatterplot. One example is given in the footnote[6].

1.3.2 Dot plots and the mean

Sometimes two variables is one too many: only one variable may be of interest. In these cases, a dot plot provides the most basic of displays. A dot plot is a one-variable scatterplot, and a sample dot plot is shown in Figure 1.12.

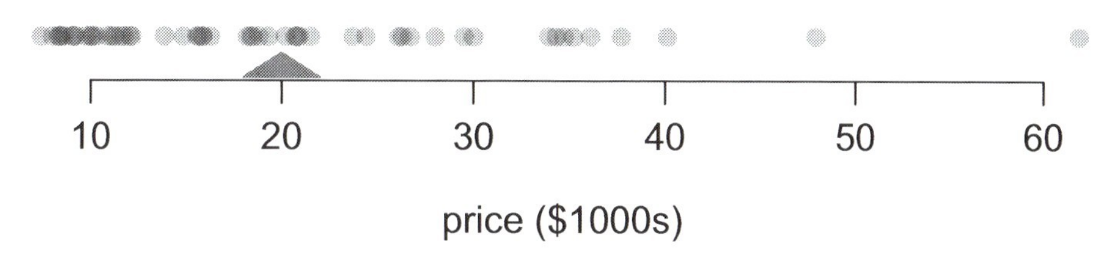

Figure 1.12: A dot plot of `price` for the `cars` data set. The triangle marks the sample's mean price.

The **mean**, sometimes called the average, is a way to measure the center of a **distribution** of data. To find the mean price of the cars in the sample, add up all the prices and divide by the number of cases.

$$\bar{x} = \frac{15.9 + 33.9 + \cdots + 26.7}{54} = 19.99259 \tag{1.1}$$

The sample mean is often labeled $\bar{x}$. The letter x is being used as an abbreviation for `price`, and the bar says it is the sample average of price. It is useful to think of the mean as the balancing point of the distribution.

$\bar{x}$
sample mean

[6]Consider the case where your vertical axis represents something "good" and your horizontal axis represents something that is only good in moderation. Health and water consumption fit this description since water becomes toxic when consumed in excessive quantities.

Mean

The sample mean of a numerical variable is computed as the sum of all of the observations divided by the number of observations:

$$\bar{x} = \frac{x_1 + x_2 + \cdots + x_n}{n} \tag{1.2}$$

where $x_1, x_2, \dots, x_n$ represent the n observed values.

⊙ **Exercise 1.8** Examine equations (1.1) and (1.2) above. What does x_1 correspond to? And x_2? Can you infer a general meaning to what x_i might represent. Answers in the footnote[7].

⊙ **Exercise 1.9** What was n in the `cars` data set? Answer in the footnote[8].

pulation
ean

The *population* mean is also computed in the same way, however, it has a special label: μ. The symbol, μ, is the Greek letter *mu* and represents the average of all observations in the population. Sometimes a subscript, such as $_x$, is used to represent which variable the population mean refers to, i.e. μ_x.

● **Example 1.3** The average price of all cars from 1993 can be estimated using the sample data. Based on the `cars` sample, what would be a reasonable estimate of μ_x, the mean price of cars from 1993?

The sample mean may provide a good estimate of μ_x. While this estimate will not be perfect, it provides a *single point estimate* of the population mean.

1.3.3 Histograms and shape

Dot plots show the exact value for each observation. This is great for small data sets, but can become problematic for larger samples. Rather than showing the value of each observation, we might prefer to think of the value as belonging to *bin*. For example, in the `cars` data set, we could create a table of counts for the number of cases with price between $5,000 and $10,000, then the number of cases between $10,000 to $15,000, and so on. Observations that fall on the boundary of a bin (e.g. $10,000) are allocated to the lower bin. This tabulation is shown in Table 1.13. To make the data easier to see visually, these binned counts are plotted as a bar chart in Figure 1.14. When the bins are all of equal width and in consecutive order, the resulting plot is called a **histogram**.

Histograms provide a view of the **data density**. Higher bars represent where the data is relatively more common. For instance, there are many more cars with a price below $15,000 than cars that cost at least $50,000 in the data set. The bars

[7] x_1 corresponds to the price of the first car (15.9), x_2 to the price of the second car (33.9), and x_i would correspond to the price of the ith car in the data set.

[8] The sample size is $n = 54$.

Price	5-10	10-15	15-20	20-25	25-30	30-35	···	55-60	60-65
Count	11	11	10	7	6	3	···	0	1

Table 1.13: The counts for the binned `price` data.

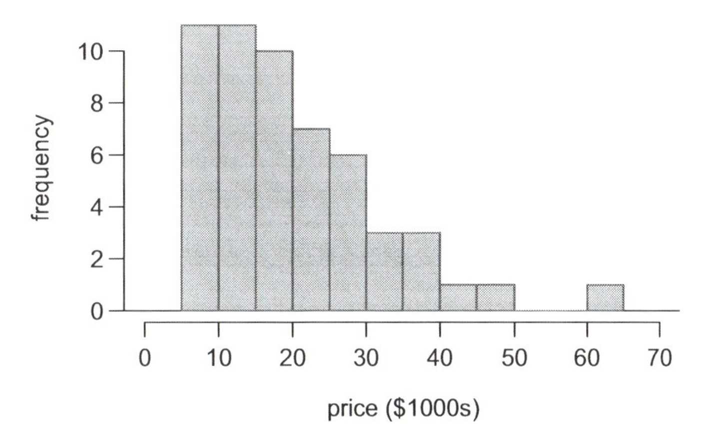

Figure 1.14: Histogram of `price`.

also make it especially easy to see how the density of the data changes from one price to another: in general, the higher the price, the fewer the cars.

Histograms are especially convenient for describing the data's **shape**. Figure 1.14 shows that most cars have a lower price, while fewer cars have higher prices. When data trails off to the right in this way and has a longer right tail, the shape is said to be **skewed to the right**[9].

Data sets with the reverse characteristic – a long, thin tail to the left – are said to be **left skewed**. It might also be said that such a distribution has a long left tail. Data sets that show roughly equal trailing off in both directions are called **symmetric**.

> **Long tails to identify skew**
> When data trails off in one direction, it is called a **long tail**. If a distribution has a long left tail, it is left skewed. If a distribution has a long right tail, it is right skewed.

⊙ **Exercise 1.10** Take a look at Figure 1.12 on page 11. Can you see the skew in the data? Is it easier to see the skew in Figure 1.12 or Figure 1.14?

⊙ **Exercise 1.11** Besides the mean (since it was labeled), what can you see in Fig-

[9]Other ways to describe data skewed to the right: **right skewed**, **skewed to the high end**, or **skewed to the positive end**.

ure 1.12 that you cannot see in 1.14? Answer in the footnote[10].

In addition to looking at whether a distribution is skewed or symmetric, histograms can be used to identify modes. A **mode is represented by a prominent peak in the distribution**[11]. There is only one prominent peak in the histogram of `price`.

Figure 1.15 shows histograms that have one, two, and three prominent peaks. Such distributions are called **unimodal**, **bimodal**, and **multimodal**, respectively. Any distribution with more than 2 prominent peaks is called multimodal. Notice that there was one prominent peak in the unimodal distribution with a second less prominent peak that was not counted since it only differs from its neighboring bins by a few observations.

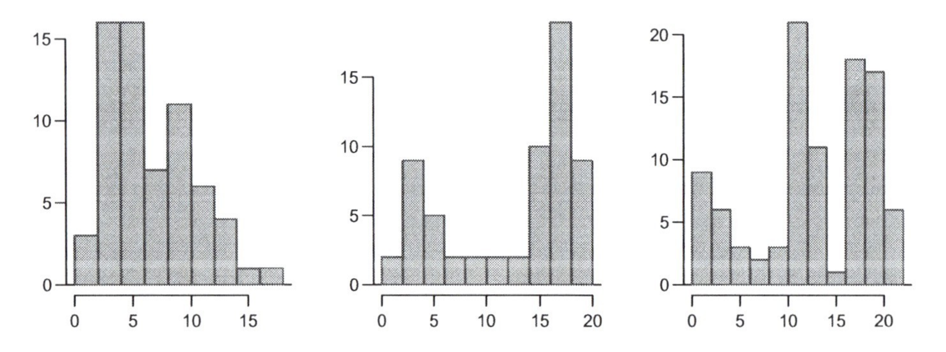

Figure 1.15: Counting only prominent peaks, the distributions are (left to right) unimodal, bimodal, and multimodal.

⊙ **Exercise 1.12** Figure 1.14 reveals only one prominent mode in `price`. Is the distribution unimodal, bimodal, or multimodal?

⊙ **Exercise 1.13** Height measurements of everyone at a K-2 elementary school were taken, which includes only young students and adult teachers. How many modes would you anticipate in this height data set? Answer in the footnote[12].

TIP: Looking for modes
Looking for modes isn't about a clear and correct answer about the number of modes in a distribution, which is why *prominent* is not defined mathematically here. The importance of this examination is to better understand your data and how it might be structured.

[10] The individual prices.

[11] Another definition of mode, which is not typically used in statistics, is the value with the most occurrences. It is common to have *no* observations with the same values in a data set, which makes this other definition useless for many real data sets.

[12] There might be two height groups visible in the data set: one of the students and one of the adults. That is, the data might be bimodal.

1.3.4 Variance and standard deviation

The mean was introduced as a method to describe the center of a data set but the data's variability is also important. Here two measures of variability are introduced: the variance and the standard deviation. Both of these are very useful in data analysis, even though their formulas are a bit tedious to compute by hand.

We call the distance of an observation from its mean its **deviation**. Below are the 1st, 2nd, and 54th deviations for the `price` variable:

$$\begin{aligned} x_1 - \bar{x} &= 15.9 - 20 = -4.1 \\ x_2 - \bar{x} &= 33.9 - 20 = 13.9 \\ &\vdots \\ x_n - \bar{x} &= 26.7 - 20 = 6.7 \end{aligned}$$

If we square these deviations and then take an average, the result is about equal to the sample **variance**, denoted by s^2:

s^2 sample variance

$$s^2 = \frac{(-4.1)^2 + (13.9)^2 + (6.7)^2}{54 - 1} = \frac{16.8 + 193.2 + 44.9}{53} = 132.4$$

We divide by $n - 1$ instead of n for reasons described in the footnote[13]. Notice that squaring the deviations does two things: (i) it makes large values much larger, seen by comparing $(-4.1)^2$, 13.9^2, and 6.7^2, and (ii) it gets rid of any negative signs.

The **standard deviation** is defined as the square root of the variance: $s = \sqrt{132.4} = 11.5$. If we like, a subscript of $_x$ may be added to the the variance and standard deviation – that is, s_x^2 and s_x – as a reminder that these are the variance and standard deviation of the observations represented by x_1, x_2, ..., x_n. These may be omitted when it is clear what the standard deviation is referencing.

s sample standard deviation

> **Variance and standard deviation**
> The variance is roughly the average squared distance from the mean. The standard deviation is the square root of the variance.

Computing the variance and standard deviation for a population uses the same formulas and methods as for a sample[14]. However, like the mean, the population values have special symbols: σ^2 for the variance and σ for the standard deviation. The symbol σ is the Greek letter *sigma*. As with the sample variance and standard deviation, subscripts such as $_x$ can be added to specify what the population variance and standard deviation represent.

σ^2 population variance

σ population standard deviation

The standard deviation is useful in considering how close the data is to the mean. Usually about 70% of the data is within one standard deviation of the

[13] The population of all cars from 1993 has some precise variance in `price`. Our estimate of this variance tends to be slightly better if we divide by $n - 1$ instead of n.

[14] However, the population variance has a division by n instead of $n - 1$.

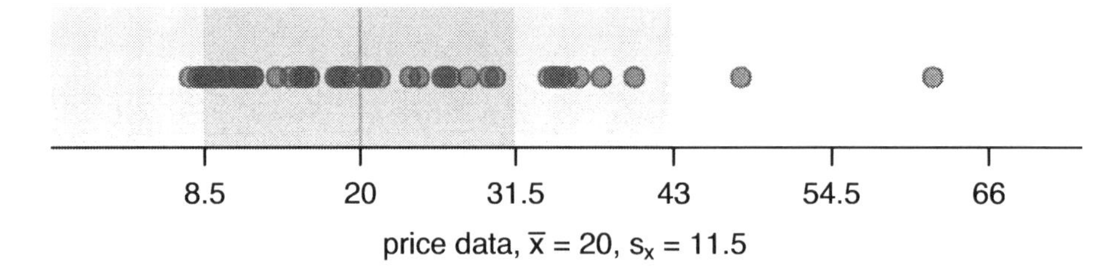

Figure 1.16: In the `price` data, 40 of 54 cars (74%) are within 1 standard deviation of the mean, $20,000. Additionally, 52 of the 54 cars (96%) and 53 of the 54 prices (98%) are within 2 and 3 standard deviations, respectively.

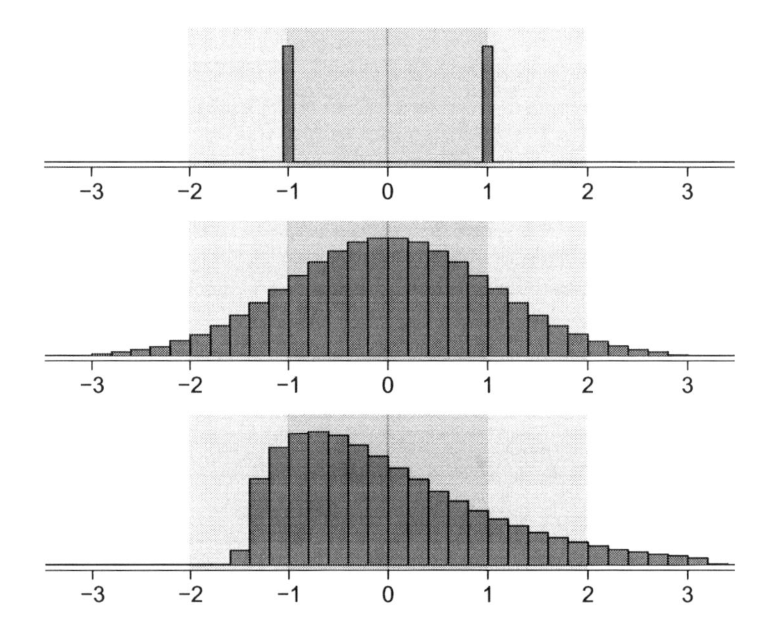

Figure 1.17: Three very different population distributions with the same mean $\mu = 0$ and standard deviation $\sigma = 1$.

mean and 95% within two standard deviations. However, these percentages can and do vary from one distribution to another. Figure 1.17 shows several different distributions that have the same center and variability but very different shapes.

⊙ **Exercise 1.14** On page 13, the concept of distribution shape was introduced. A description of the shape of a distribution should include modality and whether the distribution is symmetric or skewed to one side. Using Figure 1.17 as an example, explain why such a description is important.

● **Example 1.4** Describe the distribution of the `price` variable, shown in a histogram on page 13. The description should incorporate the center, variability, and shape of the distribution, and it should also be placed in context of the problem: the price of cars. Also note any especially unusual cases.

The distribution of car prices is unimodal and skewed to the high end. Many of the prices fall near the mean at \$20,000, and most fall within one standard deviation (\$11,500) of the mean. There is one very expensive car that costs more than \$60,000.

In practice, the variance and standard deviation are sometimes used as a means to an end, where the "end" is being able to accurately estimate the uncertainty associated with a sample's estimate. For example, in Chapter 4 we use the variance and standard deviation to assess how close the sample mean is to the population mean.

TIP: standard deviation describes variability
Standard deviation is complex mathematically. However, it is not conceptually difficult. It is useful to remember that usually about 70% of the data is within one standard deviation of the mean and about 95% is within two standard deviations.

1.3.5 Box plots, quartiles, and the median

A box plot summarizes a data set using five statistics while also plotting unusual observations. Figure 1.18 provides a vertical dot plot alongside a box plot of the `price` variable from `cars`.

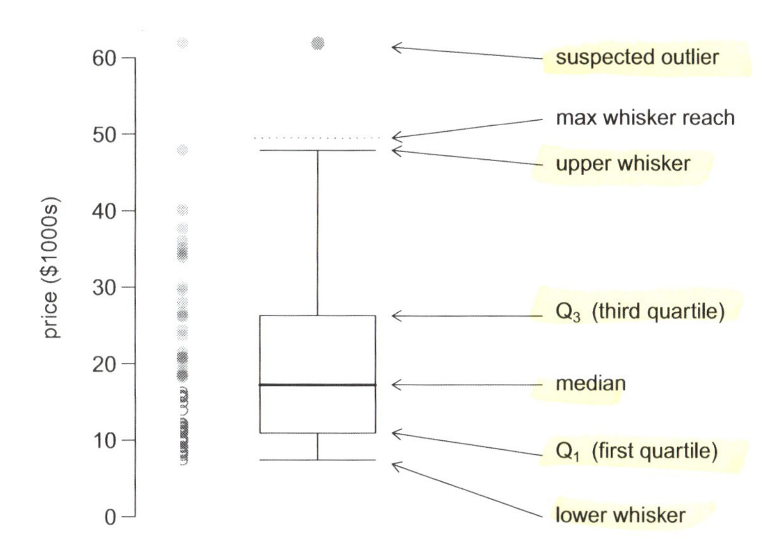

Figure 1.18: A vertical dot plot next to a labeled box plot of the `price` data. The median (\$17,250), splits the data into the bottom 50% and the top 50%, marked in the dot plot by dark-colored hollow circles and light filled circles, respectively.

The first step in building a box plot is drawing a rectangle to represent the middle 50% of the data. The total length of the box, shown vertically in Figure 1.18, is called the **interquartile range** (IQR, for short). It, like the standard deviation, is a measure of variability in data. The more variable the data, the larger the standard deviation and IQR. The two boundaries of the box are called the **first quartile** (the 25th percentile, i.e. 25% of the data falls below this value) and the **third quartile** (the 75th percentile), and these are often labeled Q_1 and Q_3, respectively.

Interquartile range (IQR)
The **interquartile range (IQR)** is the length of the box in box plot. It is computed as

$$IQR = Q_3 - Q_1$$

where Q_1 and Q_3 are the 25th and 75th percentiles.

The line splitting the box marks the **median**, or the value that splits the data in half. Figure 1.18 shows 50% of the data falling below the median (dark hollow circles) and other 50% falling above the median (light-colored filled circles). There are 54 car prices in the data set so the data is perfectly split into two groups. We take the median in this case to be the average of the two observations closest to the 50th percentile: $\frac{\$16{,}300+\$18{,}200}{2} = \$17,250$. When there are an odd number of observations, there will be exactly one observation that splits the data into two halves, and in this case that observation is the median (no average needed).

Median: the number in the middle
If the data was ordered from smallest to largest, the **median** would be the observation right in the middle. If there are an even number of observations, there will be two values in the middle, and the median is taken as their average.

⊙ **Exercise 1.15** How much of the data falls between Q_1 and the median? How much between the median and Q_3? Answers in the footnote[15].

Extending out from the box, the **whiskers** attempt to capture the data outside of the box, however, their maximum reach is only $1.5*IQR$.[16] They grab everything within this reach. In Figure 1.18, the upper whisker cannot extend to the last point, which is beyond $Q_3 + 1.5 * IQR$, and so extends only to the last point below this limit. The lower whisker stops at the lowest price, $7,400, since there is no

[15]Since Q_1 and Q_3 capture the middle 50% of the data and the median splits the middle of the data, 25% of the data falls between Q_1 and the median, and another 25% falls between the median and Q_3.

[16]While the choice of exactly 1.5 is arbitrary, it is the most commonly used value for box plots.

additional data to reach. In a sense, the box is like the body of the box plot and the whiskers are like its arms trying to reach the rest of the data.

Any observation that lies beyond the whiskers is labeled with a dot. The purpose of labeling these points – instead of just extending the whiskers to the minimum and maximum observed values – is to help identify any observations that appear to be unusually distant from the rest of the data. Unusually distant observations are called **outliers**. In the case of the `price` variable, the car with price $61,900 is a potential outlier.

Outliers are extreme
An **outlier** is an observation that appears extreme relative to the rest of the data.

Examination of data for possible outliers serves many useful purposes, including

1. identifying extreme skew in the distribution.
2. finding evidence that extreme cases are common for the variable being examined.
3. identifying data collection or entry errors. If there was car price listed as $140,000, it is worth reviewing the observation to see whether it was really $14,000.
4. providing insights into interesting phenomena with the data.

TIP: Why it is important to look for outliers
The identification of outliers is not actually what is important, which is why no rigid definition for outlier is provided. What is important is to examine and take special note of *suspected* outliers and why they are extreme from the rest of the data.

⊙ **Exercise 1.16** The observation $61,900, a suspected outlier, was found to be an accurate observation. What would such an observation suggest about the nature of vehicle prices?

⊙ **Exercise 1.17** Using Figure 1.18 on page 17, estimate the following values for `price` in the `cars` data set: (a) Q_1, (b) Q_3, (c) IQR, and (d) the maximum length of the upper whisker. How are (c) and (d) related?

1.3.6 Robust statistics

How would sample statistics be affected if $200,000 was observed instead of $61,900 for the most expensive car? Or what if $200,000 had been in the sample instead

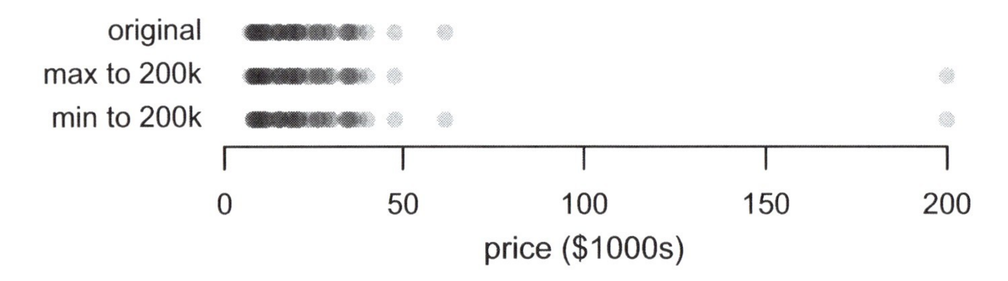

Figure 1.19: Dot plots of the original price data and two modified price data sets.

	robust		not robust	
scenario	median	IQR	$\bar{x}$	s
original `price` data	17.25	15.30	19.99	11.51
move \$61,900 to \$200,000	17.25	15.30	22.55	26.53
move \$7,400 to \$200,000	18.30	15.45	26.12	35.79

Table 1.20: A comparison of how the median, IQR, mean ($\bar{x}$), and standard deviation (s) change when extreme observations are in play.

of the cheapest car at \$7,400? These two scenarios are plotted alongside the original data in Figure 1.19, and sample statistics are computed under each of these scenarios in Table 1.20.

⊙ **Exercise 1.18** (a) Which is more affected by extreme observations, the mean or median? Table 1.20 may be helpful. (b) Is the standard deviation or IQR more affected by extreme observations?

The median and IQR are called **robust estimates** because extreme observations have little effect on their values. The mean and standard deviation are much more affected by changes in extreme observations.

⊙ **Exercise 1.19** Why doesn't the median or IQR change from the original `price` data to the second scenario of Table 1.20?

⊙ **Exercise 1.20** Why are robust statistics useful? If you were searching for a new car and cared about price, would you be more interested in the mean vehicle price or the median vehicle price when deciding what price you should pay for a regular car?

1.4 Considering categorical data

Like numerical data, categorical data can also be organized and analyzed. In this section, tables and other basic tools for categorical data analysis are introduced that will be used throughout this book.

1.4.1 Contingency tables

Table 1.21 summarizes two variables from the `cars` data set: `type` and `driveTrain`. A table that summarizes data for two categorical variables in this way is called a **contingency table**. Each number in the table represents the number of times a particular combination of variable outcomes occurred. For example, the number 19 corresponds to the number of cars in the data set that are small *and* have front wheel drive. Row and column totals are also included. The **row totals** equal the total counts across each row (e.g. $19 + 0 + 2 = 21$), and **column totals** are total counts down each column.

	front	rear	4WD	total
small	19	0	2	21
midsize	17	5	0	22
large	7	4	0	11
total	43	9	2	54

Table 1.21: A contingency table for `type` and `driveTrain`.

A table for a single variable is called a **frequency table**. Table 1.22 is a frequency table for the `type` variable. If we replaced the counts with percentages or proportions, the table would be called a **relative frequency table**.

small	midsize	large
21	22	11

Table 1.22: A frequency table for the `type` variable.

⊙ **Exercise 1.21** Examine Tables 1.21 and 1.22. Why is Table 1.22 redundant if Table 1.21 is provided?

1.4.2 Bar charts and proportions

A bar chart is a common way to display a single categorical variable. The left panel of Figure 1.23 shows a **bar plot** of `type`. In the right panel, the counts were converted into proportions (e.g. $21/54 = 0.389$ for `small`), making it easy to compare how often different outcomes occur irrespective of sample size.

⊙ **Exercise 1.22** Which of the following statements would be more useful to an auto executive? (1) 21 cars in our sample were `small` vehicles. (2) 38.9% of the cars in

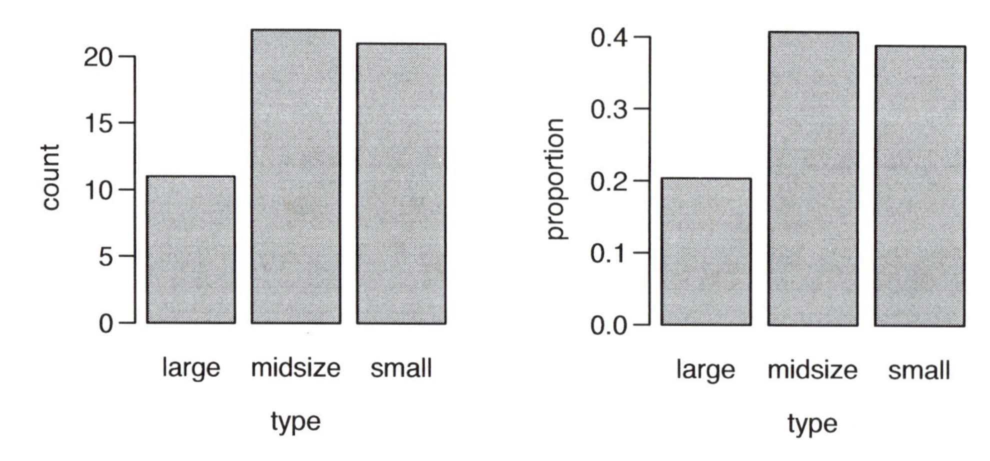

Figure 1.23: Two bar plots of `type`. The left panel shows the counts and the right panel the proportions in each group.

our sample were `small` vehicles. Comment in the footnote[17].

If proportions are so great, why not just change Table 1.21 into a table of proportions? Because there is a better way.

Table 1.24 shows the row proportions for Table 1.21. The **row proportions** are computed as the counts divided by their row totals. The count 17 at the intersection of `midsize` and `front` is replaced by $17/22 = 0.773$, i.e. 17 divided by its row total, 22. So what does 0.773 represent? It corresponds to the proportion of midsize vehicles in the sample that have front wheel drive.

	`front`	`rear`	`4WD`	total
`small`	$19/21 = 0.905$	$0/21 = 0.000$	$2/21 = 0.095$	1.000
`midsize`	$17/22 = 0.773$	$5/22 = 0.227$	$0/22 = 0.000$	1.000
`large`	$7/11 = 0.636$	$4/11 = 0.364$	$0/11 = 0.000$	1.000
total	$43/54 = 0.796$	$9/54 = 0.167$	$2/54 = 0.037$	1.000

Table 1.24: A contingency table with row proportions for the `type` and `driveTrain` variables.

A contingency table of the column proportions is computed in a similar way, where each **column proportion** is computed as the count divided by the corresponding column total. Table 1.25 shows such a table, and here the value 0.442 represents the proportion of front wheel drive cars in the sample that that are small cars.

[17] Even if the sample size (54) was provided in the first statement, the auto exec would probably just be trying to figure out the proportion in her head.

	front	rear	4WD	total
small	$19/43 = 0.442$	$0/9 = 0.000$	$2/2 = 1.000$	$21/54 = 0.389$
midsize	$17/43 = 0.395$	$5/9 = 0.556$	$0/2 = 0.000$	$22/54 = 0.407$
large	$7/43 = 0.163$	$4/9 = 0.444$	$0/2 = 0.000$	$11/54 = 0.204$
total	1.000	1.000	1.000	1.000

Table 1.25: A contingency table with column proportions for the `type` and `driveTrain` variables.

⊙ **Exercise 1.23** What does 0.364 represent in Table 1.24? Answer in the footnote[18]. What does 0.444 represent in Table 1.25?

⊙ **Exercise 1.24** What does 0.796 represent in Table 1.24? Answer in the footnote[19]. What does 0.407 represent in the Table 1.25?

● **Example 1.5** Researchers suspect the proportion of male possums might change by location. A contingency table for the `pop` (living location) and `sex` variables from the `possum` data set is shown in Table 1.26. Based on these researchers' interests, which would be more appropriate: row or column proportions?

The interest lies in how the `sex` changes based on `pop`. This corresponds to the row proportions: the proportion of males/females in each location.

	f	m	total
Vic	24	22	46
other	19	39	58
total	43	61	104

Table 1.26: A contingency table for `pop` and `sex` from the `possum` data set.

Example 1.5 points out that row and column proportions are not created equal. It is important to consider each before settling on one to ensure that the most useful table is constructed.

1.4.3 Mosaic plots and independence

Contingency tables using row or column proportions are especially useful for examining how two categorical variables are related. Mosaic plots provide a way to put these tables into a graphical form. To reduce complexity, this section we will only consider vehicles with front and rear wheel drive, as shown in Tables 1.27(a) and 1.27(b).

[18] 0.364 represents the proportion of large cars in the sample that are rear wheel drive.
[19] 0.827 represents the proportion of cars in the sample that are front wheel drive vehicles.

	front	rear	total
small	19	0	19
midsize	17	5	22
large	7	4	11
total	43	9	52

(a)

	front	rear
small	1.00	0.00
midsize	0.77	0.23
large	0.64	0.36
total	0.83	0.17

(b)

Table 1.27: (a) Contingency table for `type` and `driveTrain` where the two vehicles with `driveTrain` = `4WD` have been removed. (b) Row proportions for Table (a).

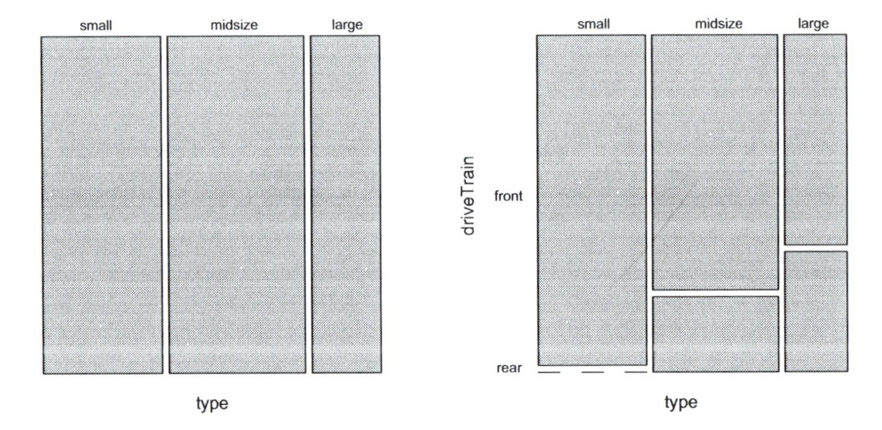

Figure 1.28: The one-variable mosaic plot for `type` and the two-variable mosaic plot for both `type` and `driveTrain`.

A **mosaic plot** is a graphically display of contingency table information. Here we construct the mosaic plot representing the row proportions of `type` and `driveTrain` in Table 1.27(b).

The left panel of Figure 1.28 shows a mosaic plot for the `type` variable. Each column represents a level of `type`, and the column widths corresponds to the proportion of cars of each type. For instance, there are fewer small cars than midsize cars, so the small car column is slimmer.

This plot is further broken into pieces in the right panel using the `driveTrain` variable. Each column is split proportionally according to the drivetrain of the vehicles of that particular type. For example, the first column representing only small cars was broken into small cars with front and rear drivetrains. As another example, the top of the middle column represents midsize cars with front wheel drive, and the lower part of the middle column represents midsize cars with rear wheel drive. Because each column is broken apart in very different places, this suggests the proportion of vehicles with front wheel drive differs with vehicle `type`. That is, `driveTrain` and `type` show some connection and are therefore associated.

In a similar way, a mosaic plot representing column proportions of Table 1.21

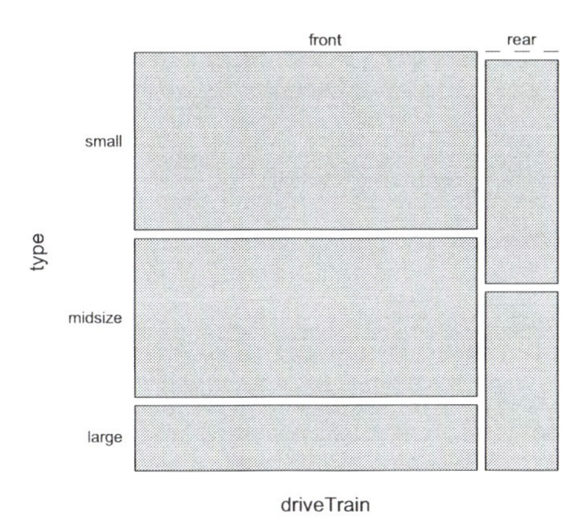

Figure 1.29: Mosaic plot where type is broken up within the drivetrain.

can be constructed, which is shown in Figure 1.29.

⊙ **Exercise 1.25** Why is it that the combination `rear-small` does not have an actual rectangle? Hint in the footnote[20].

⊙ **Exercise 1.26** Describe how the mosaic plot shown in Figure 1.29 was constructed. Answer in the footnote[21].

1.4.4 The only pie chart you will see in this book

While pie charts are well known, they do not typically do as good of a job as other charts in a data analysis. A **pie chart** is shown in Figure 1.30 on the next page along side a bar plot. It is more difficult to compare group sizes in a pie chart than in a bar plot. While pie charts may be helpful in some scenarios, they are not typically as helpful in a strict data analysis setting, which is why this is the first and last pie chart in this book.

⊙ **Exercise 1.27** Using the pie chart, is it easy to tell which level, `midsize` or `small`, has a larger proportion in the sample? What about when using the bar plot?

1.4.5 Comparing numerical data across groups

Some of the more interesting investigations can be considered by examining numerical data across groups. The methods required aren't really new. All that is required is to make a numerical plot for each group. Here two convenient methods are introduced: side-by-side box plots and hollow histograms.

[20]How many cars have this combination in Table 1.27?

[21]First, the cars were split up by `driveTrain` into two groups represented by the columns. Then the `type` variable splits each of these columns into the levels of `type`.

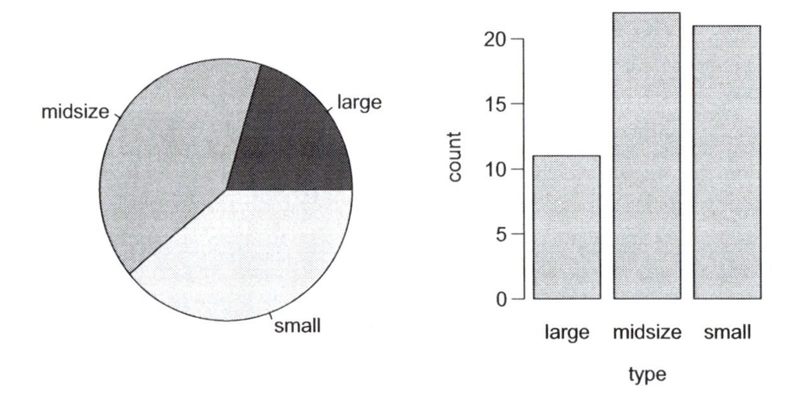

Figure 1.30: A pie chart and bar plot of `type` for the data set `cars`.

small		**midsize**		**large**
15900	11600	33900	28000	20800
9200	10300	37700	35200	23700
11300	11800	30000	34300	34700
12200	9000	15700	61900	18800
7400	11100	26300	14900	18400
10100	8400	40100	26100	29500
8400	10900	15900	21500	19300
12100	8600	15600	16300	20900
8000	9800	20200	18500	36100
10000	9100	13900	18200	20700
8300		47900	26700	24400

Table 1.31: The data from the `price` variable split up by `type`.

From the data set `cars`, we would like to compare vehicle price according to vehicle type. There are three levels of `type` (`small`, `midsize`, and `large`), and the vehicle prices can be split into each of these groups, as shown in Table 1.31.

The **side-by-side box plot** is a traditional tool for comparing across groups, and it is shown in the left panel of Figure 1.32. This is just three box plots – one for each `type` – placed into one plotting window and drawn on the same scale.

Another plotting method worthy of mention is **hollow histograms**, which are just the histogram outlines of each group put on the same plot, as shown in the right panel of Figure 1.32.

⊙ **Exercise 1.28** Use each plot in Figure 1.32 to compare the vehicle prices across groups. What do you notice about the approximate center of each group? What do you notice about the variability between groups? Is the shape relatively consistent between groups? How many *prominent* modes are there for each group?

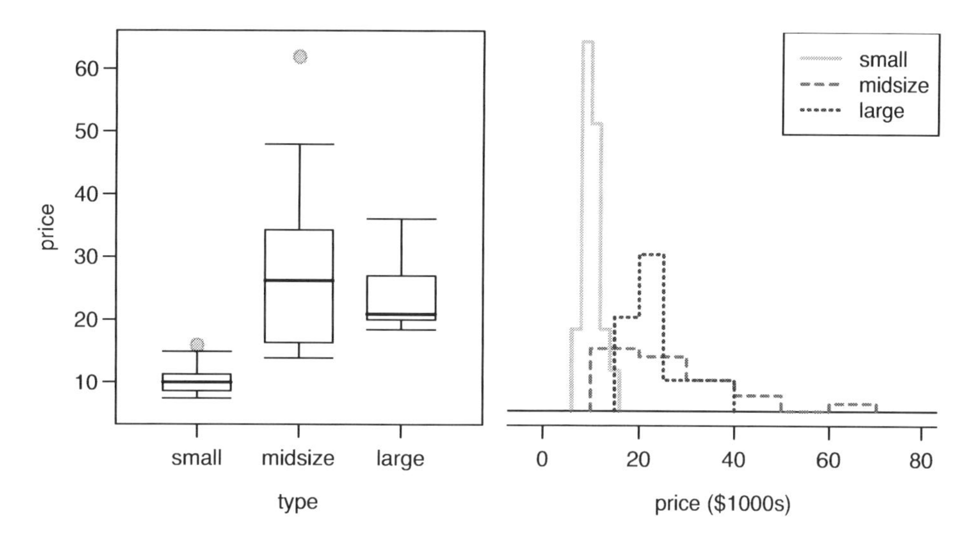

Figure 1.32: Side-by-side box plot (left panel) and hollow histograms (right panel) for `price` where the groups are each level of `type`.

⊙ **Exercise 1.29** What components of each plot in Figure 1.32 do you find most useful?

1.5 Data collection

The first step in conducting research is to identify topics or questions that are to be investigated. A clearly laid out research question is helpful in identifying what subjects or cases are to be studied and what variables are important. This information provides a foundation for *what* data will be helpful. It is also important that we consider *how* data is collected so that it is trustworthy and helps achieve the research goals. This section outlines important ideas and practices in data collection.

1.5.1 Populations and samples

Consider the following three research questions:

1. What is the average mercury content in swordfish in the Atlantic Ocean?

2. Over the last 5 years, what is the average time to degree for UCLA undergraduate students?

3. Does the drug sulphinpyrazone reduce the number of deaths in heart attack patients?

In each research question, some population of cases is considered. In the first question, all swordfish in the Atlantic ocean are relevant to answering the question.

Each fish represents a case, and all of these fish represent the **population** of cases. Often times, it is too expensive to collect data for every case in a population. Instead a *sample* is taken of the population. A **sample** represents a subset of the cases and often times represents only a small fraction of all cases in the population. For instance, 60 swordfish (or some other number) in the population might be selected, and this sample data may be used to provide an estimate of the population average, i.e. an answer to the research question.

⊙ **Exercise 1.30** For the second and third questions above, identify what is an individual case and also what is the population under consideration. Answers in the footnote[22].

1.5.2 Anecdotal evidence

We posed three research questions in Section 1.5.1. Below are some statements by folks who are responding to the research questions:

1. A man on the news got mercury poisoning from eating swordfish, so the average mercury concentration in swordfish must be dangerously high.
2. I met two students who took more than 10 years to graduate from UCLA, so it must take longer to graduate at UCLA than at many other colleges.
3. My friend's dad had a heart attack and died after they gave him sulphinpyrazone. The drug must not work.

Each of the conclusions made are based on some data. However, there are two problems. First, the data described only represents a one or two cases. Second and more importantly, it is unclear whether these cases are actually representative of the population. Data collected in this haphazard fashion is called **anecdotal evidence**.

Anecdotal evidence
Data collected in a haphazard fashion. Such evidence may be true and verifiable but often times represent extraordinary cases.

Anecdotal evidence typically is composed of unusual cases that we recall based on their striking characteristics. For instance, we are more likely to remember the two folks we met who took 10 years to graduate than the six others who graduated in four years.

[22](2) First, notice that this question is only relevant to students who complete their degree; the average cannot be computed using a student who never finished her degree. Thus, only UCLA undergraduate students who have graduated in the last five years represent cases in the population under consideration. (3) A heart attack patient represents a case. The population represents all heart attack patients.

Figure 1.33: In February 2010, some media pundits cited one large snow storm as valid evidence against global warming. As comedian Jon Stewart pointed out, "It's one storm, in one region, of one country."[a]

[a]www.openintro.org/clip/anecdotal.php

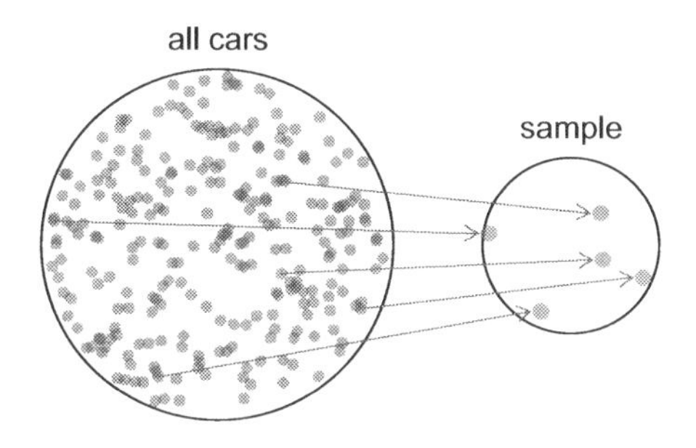

Figure 1.34: Cars from the population are randomly selected to be included in the sample.

Instead of looking at the most unusual cases, we often want to look at a sample that is representative of the population. It is also helpful to collect a sample with many cases.

1.5.3 Sampling from a population

The `cars` data set represents a sample of cars from 1993. All cars from 1993 represent the population, and the cars in the sample were **randomly** selected from the population. Randomly selected in this context is equivalent to how raffles are run. The name of each car from the population was written on a raffle ticket, and 54 tickets were drawn.

Why pick a sample randomly? Why not just pick a sample by hand? Consider the following scenario.

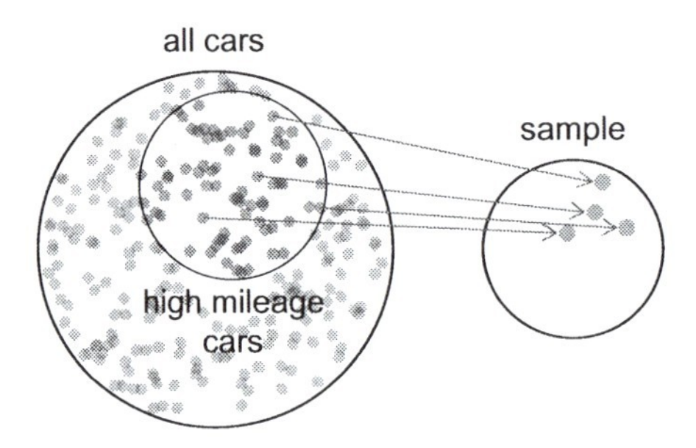

Figure 1.35: Instead of sampling from all cars from 1993, an environmentalist might inadvertently pick cars with high mileage disproportionally often.

⊙ **Exercise 1.31** Suppose a muscle car enthusiast is asked to select several cars for a study. What kind of cars do you think she might collect? Do you think her sample would be representative of all cars?

Even If someone was permitted to pick and choose exactly which cars were included in the sample, it is entirely possible that the sample could be skewed to that person's interests, which may be entirely unintentional. This introduces **bias** into a sample. Sampling randomly helps resolve this problem. The most basic random sample is called a **simple random sample**, and is the equivalent of using a raffle to select cases. This means that each case in the population has an equal chance of being included and there is no implied connection between the cases in the sample. The act of taking a simple random sample helps eliminate bias, however, it can still crop up in other ways.

Even when people are seemingly picked at random (for surveys, etc.), caution must be exercised if the **non-response** is high. For instance, if only 15% of the people randomly sampled for a survey actually respond, then it is unclear whether the results are **representative** of the entire population. **Non-response bias** can skew results one way or another.

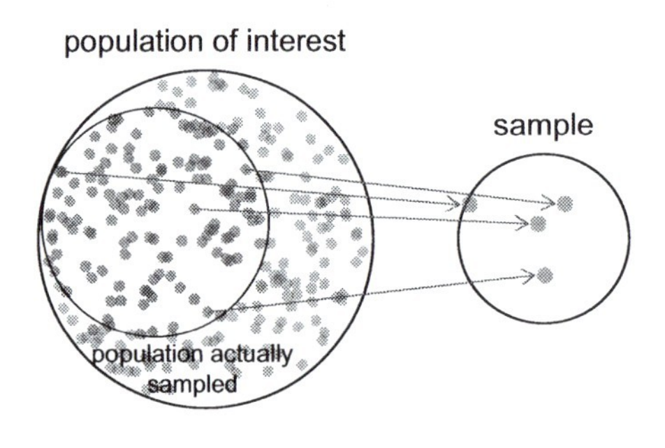

Figure 1.36: Surveys may result in only reaching a certain group within the population, and it is not obvious how to fix this problem.

Another common downfall is a **convenience sample**, where individuals who are easily accessible are more likely to be included in the sample. For instance, if a political survey is done by stopping people walking in the Bronx, this probably will not fairly represent all of New York City. It is often difficult to discern what sub-population a convenience sample represents.

⊙ **Exercise 1.32** We can easily access ratings for products, sellers, and companies through websites. These ratings are based only on those people who go out of their way to provide a rating. If a seller has a rating of 95% on Amazon, do you think this number might be artificially low or high? Why?

1.5.4 Explanatory and response variables

Consider the second question from page 6 for the `possum` data set:

(2) Will males or females, on the average, be longer?

This question might stem from the belief that a possum's gender might in some way affect its size but not the reverse. If we suspect a possum's sex might affect its total length, then `sex` is the **explanatory** variable and `totalL` is the **response** variable in the relationship[23]. If there are many variables, it may be possible to label a number of them as explanatory and the others as response variables.

TIP: Explanatory and response variables
To identify the explanatory variable in a pair of variables, identify which of the two is suspected of affecting the other.

explanatory variable —might affect→ response variable

Caution: association does not imply causation
Labeling variables as *explanatory* and *response* does not guarantee the relationship between the two is actually causal, even if there is an association identified between the two variables. We use these labels only to keep track of which variable we suspect affects the other.

In some cases, there is no explanatory or response variable. Consider the first question from page 6:

(1) If a possum has a shorter-than-average head, do you think its skull width will be smaller or larger than the average skull width?

This question does not have an explanatory variable since it doesn't really make sense that `headL` would affect `skullW` or vice-versa, i.e. the direction is ambiguous.

[23] Sometimes the explanatory variable is called the **independent** variable and the response variable is called the **dependent** variable. However, this becomes confusing since a *pair* of variables might be independent or dependent, so we avoid this language.

1.5.5 Experiments

An association between two variables means they are in some way connected. It would be useful to expand on this connection and say that one variable causes the other to change, however, it is not clear that is always the case. When is such a conclusion reasonable?

Consider the variables `pop` (living location) and `sex` from the `possum` data set, summarized in Table 1.37 using row proportions. The data suggests there might be a difference: 48% of the sample was male for those possums from Victoria while 67% of the sample was male for the other locations. Suppose these proportions were the same in the population as in the sample, i.e. that there was a real difference in gender ratio in the two geographical regions[24].

	female	male
Victoria	0.52	0.48
other	0.33	0.67

Table 1.37: Contingency table with row proportions of possums' sex based on location.

⊙ **Exercise 1.33** If we suspect that the living location affects the proportion of males in the possum population, which variable would be the explanatory variable and which the response?

Does Exercise 1.33 imply that living in New South Wales or Queensland *causes* a possum to be male more often than it would in Victoria? Put it another way: if possums were transported from Victoria to New South Wales or Queensland, does this actually mean their offspring is more likely to be male than if it had been born in Victoria? The data doesn't answer this *causal* question.

To answer whether living location does affect gender, it would be interesting to take a bunch of possums and try it out:

- Collect a random sample of 100 possums from Victoria (or as best a random sample as is possible).
- Randomly split the 100 possums into two groups. Leave one group in Victoria and transport the other group to New South Wales and Queensland.
- Observe the offspring of each group.

This study is called an **experiment** because the possum locations – the explanatory variable – are not simply observed but are imposed on the possums by the

[24]The biological mechanism for gender suggests the ratio should always be one male to one female. However, the animals that actually survive might not do so in a one-to-one ratio and other factors may also play into the resulting proportion of males in a population.

researchers in this setup. Furthermore, the group assignments are randomly generated through the equivalent of a raffle, making this a **randomized experiment**.

If there is a big difference in these two groups again in the experiment, then this new data would provide strong evidence that living in New South Wales and Queensland actually *causes* more male possums to be born and survive. If that difference seems to disappear or be minor, then living in New South Wales and Queensland may not cause more male possums to survive to adulthood but may just be *associated* with more males surviving. That is, it might be true that the variables really are associated but they may not share a causal relationship.

TIP: association $\neq$ causation
In general, association does not imply causation, and causation can only be inferred from an experiment.

TIP: Three principles of designing experiments

Control. Control the value of the explanatory variable in each case, and control the effects of other potentially influential variables.

Randomize. Use **subjects** (i.e. cases) from a random sample of the population when possible. Subjects should be randomly split into groups, where groups are based on the explanatory variable.

Replicate. Observe many cases or run many trials.

Controlling variables that might affect both the explanatory variable and response may be the most difficult part of an experiment because such variables are difficult to identify. For instance, it would be appropriate to control for (or take into account) which possums are siblings in an analysis of the hypothetical possum experiment. In an experiment with human subjects[25], there are complex factors to control for, and these will be discussed in more detail in Section 1.5.7.

If study subjects are not from a random sample of the population, additional justification is needed to extend an experiment's conclusions to the population.

● **Example 1.6** Describe how the three experimental principles fit into the hypothetical possum experiment.

Control: the location, representing the explanatory variable, would be assigned to each possum. Randomize: the possums would be randomly split into two groups, where one group goes to Victoria and the other group goes to New South Wales or Queensland. Replicate: include many (100) possums in the study.

[25]Human subjects are more often called **patients**, **volunteers**, or **study participants**.

1.5.6 Observational studies

The `possum` data set was actually from an **observational study** since the researchers did not randomly assign which possums lived where. Generally, data in observational studies is collected only by monitoring what occurs while experiments require the explanatory variable to be assigned for each subject by the researchers.

Inferring causal conclusions from experiments is often reasonable. However, making the same causal conclusions from observational data can be treacherous and is not recommended. Thus, we can generally only infer associations from observational data.

⊙ **Exercise 1.34** Suppose an observational study tracked sunscreen use and skin cancer and it was found that the more sunscreen someone used, the more likely they were to have skin cancer (!). Does this mean sunscreen *causes* skin cancer?

Previous research tells us that using sunscreen actually reduces skin cancer risk, so maybe there is another variable that can explain this apparent association between sunscreen usage and skin cancer. One important piece of information absent is sun exposure. If someone is out in the sun all day, she is more likely to use sunscreen *and* more likely to get skin cancer.

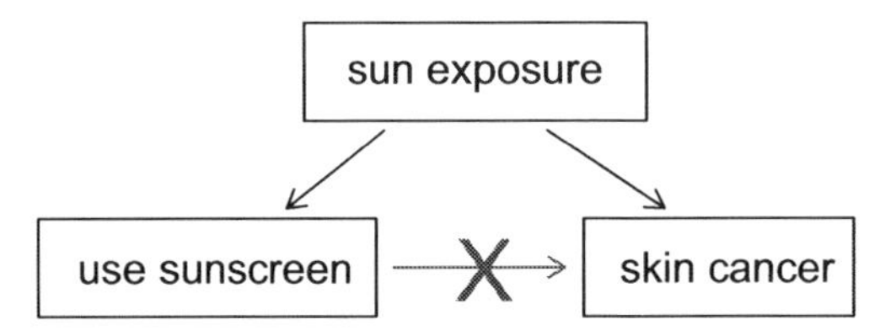

It just so happens that if someone is exposed to the sun they also usually use sunscreen. Exposure to the sun is unaccounted for in the investigation, giving the incorrect impression that sunscreen causes skin cancer.

Sun exposure is what is called a **lurking variable**, which is a variable that is the true cause for change in the response. While one method to justify making causal conclusions from observational studies is to exhaust the search for lurking variables, there is no guarantee that all lurking variables can be examined or measured.

In the same way, the `possum` data set is an observational study with possible lurking variables of its own, and its data cannot easily be used to make causal conclusions.

⊙ **Exercise 1.35** There appears to be a real difference in the proportion of possums that are male based on location. However, it is unreasonable to conclude that this is a causal relationship because the data is observational. Suggest at least one lurking variable that might be the true cause for the differences in `sex`. One lurking variable is listed in the footnote[26].

[26]Some genes can affect one gender more than the other. If the `other` population has a gene

1.5.7 Reducing bias in human experiments

Randomized experiments are the gold standard for data collection but do not ensure an unbiased perspective into the cause and effect relationships in all cases. Human studies are perfect examples where bias can unintentionally arise. Here we reconsider the sulphinpyrazone study, which was described in Section 1.1.

Researchers wanted to examine whether a drug called sulphinpyrazone would reduce the number of deaths after heart attacks. They designed an experiment because they wanted to draw causal conclusions about the drug's effect. Study volunteers were randomly placed into two study groups. One group, the **treatment group**, received the drug. The other group, called the **control group**, did not receive any drug treatment.

Put yourself in the place of a person in the study. If you are in the treatment group, you are given a fancy new drug that you anticipate will help you. On the other hand, a person in the other group doesn't receive the drug and sits idly, hoping her participation doesn't increase her risk of death. These perspectives suggest there are actually two effects: the one of interest is the effectiveness of the drug and the second is an emotional effect that is difficult to quantify.

Researchers aren't interested in this emotional effect, which might bias the study. To circumvent this problem, researchers do not want patients to know which group they are in. When researchers keep the patients in the dark about their treatment, the study is said to be **blind**. But there is one problem: if a patient doesn't receive a treatment, she will know she is in the control group. The solution to this problem is to give fake treatments to patients in the control group. A fake treatment is called a **placebo**, and an effective placebo is the key to making a study truly blind. A classic example of a placebo is a sugar pill that is made to look like the actual treatment pill. Often times, a placebo results in a slight but real improvement in patients. This often positive effect has been dubbed the **placebo effect**.

The patients are not the only ones who should be blinded: doctors and researchers can accidentally bias a study. When a doctor knows a patient has been given the real treatment, she might inadvertently give that patient more attention or care than a patient that she knows is on the placebo. To guard against this bias (which again has been found to have a measurable effect in some instances), most modern studies employ a **double-blind** setup where doctors or researchers who interact with patients are, just like the patients, unaware of who is or is not receiving the treatment[27].

that affects males more positively than females and this gene is less common in the `Vic` population, this might explain the difference in gender ratio for each level of `pop`.

[27] There are always some researchers involved in the study who do know which patients are receiving which treatment. However, they do not have interactions with the patients and do not tell the blinded doctors who is receiving which treatment.

1.5.8 Variability within data

The study examining the effect of sulphinpyrazone was double-blinded, and the results are summarized in Table 1.38. The variables have been called `group` and `outcome`. Do these results mean the drug was effective at reducing deaths? In the observed groups, a smaller proportion of individuals died in the treatment group than the control group (0.056 versus 0.081), however, it is unclear whether that difference is *convincing evidence* that the drug is effective.

		outcome		
		lived	died	Total
group	treatment	692	41	733
	control	682	60	742
	Total	1374	101	1475

Table 1.38: Summary results for the sulphinpyrazone study.

- **Example 1.7** Suppose there had been only 45 deaths in the control group and still only 41 deaths in the treatment group. Would this be convincing evidence that the drug was effective?

  The proportion is still in favor of the drug (0.056 versus 0.061). However, this sample difference seems more likely to happen than 0.056 and 0.081. It would actually be surprising if the sample proportions in each group were *exactly* equal. When we collect data, there is usually a little bit of wiggle in the data. That is, the sample data only provides an *estimate* of the truth.

Example 1.7 is a reminder that the sample will not perfectly reflect the population. It is possible to see a small difference *by chance*. Small differences in large samples can be important and meaningful but it is unclear when we should say that a difference is so large it was probably not due to chance. In Section 1.6, we evaluate whether Table 1.38 shows convincing evidence that sulphinpyrazone is effective at reducing deaths or whether we remain unconvinced.

1.6 Case study: efficacy of sulphinpyrazone*

An asterisk (*) attached to a section means that section is optional. To utilize this optional section, it is recommended that an accompanying lab using these same methods on a second data set is incorporated into the class.

- **Example 1.8** Suppose your professor splits the students in class into two groups: students on the left and students on the right. If $\hat{p}_L$ and $\hat{p}_R$ represent the proportion of students who own an Apple computer on the left and right, respectively, would you be surprised if $\hat{p}_L$ did not exactly equal $\hat{p}_R$?

While the proportions would probably be close to each other, it would be unusual for them to be exactly the same. We would probably observe a small difference due to chance.

⊙ **Exercise 1.36** If we don't think the side of the room a person sits on is related to owning an Apple computer, what assumption are we making about the relationship of the variables `height` and `appleComputer`? Answer in the footnote[28].

Table 1.38 shows there were 19 fewer deaths in the treatment group than in the control group for the sulphinpyrazone study, a difference in death rates of 2.5% $\left(\frac{60}{742} - \frac{41}{733} = 0.025\right)$. Might this difference just be due to chance? Or is this convincing evidence that sulphinpyrazone works? We label these two competing claims, H_0 and H_A:

H_0 **Independence model.** The variables `group` and `outcome` are independent. They have no relationship, and the difference in death rates, 2.5%, was due to chance.

H_A **Alternate model.** The `group` and `outcome` variables are *not* independent. The difference in death rates of 2.5% was not due to chance and the treatment did reduce the death rate.

Consider what it would mean to the study if the independence model, which says that the variables `group` and `outcome` are unrelated, is true. Each person was either going to live or die, and the drug had no effect on the outcome. The researchers were just randomly splitting up these individuals into two groups, just like we split the class in half in Example 1.8. The researchers observed a difference of 2.5% by chance.

Consider the alternative: the treatment affected the outcome. We would expect to see some difference in the groups, with a lower percentage of deaths in the group of patients who received the drug.

If the data conflicts so much with H_0 that the independence model cannot be deemed reasonable, we will reject it in favor the alternate model, H_A. In other words, we will not reject the position that H_0 is true unless the evidence from the study in favor of H_A is too convincing.

1.6.1 Simulating the study*

Suppose H_0 is true. The researchers had split the patients in the study randomly into the two groups and observed the difference. Under this model, the 2.5% difference was due to chance.

We can simulate differences due to chance using a **randomization technique**. If we randomly assign each person in the study into either a fake treatment or fake control group, then this newly assigned variable representing fake groups and the

[28]We would be assuming the variables `height` and `appleComputer` are independent.

outcome are independent. Any difference we would observe in these fake groups would be due to chance.

We run this **simulation** by taking 733 `treatmentFake` and 742 `controlFake` labels and randomly assign them to the patients[29]. We use a computer program to randomly assign these labels to the patients, and we organize these results into Table 1.39.

		outcome		
		lived	died	Total
groupFake	treatmentFake	686	47	733
	controlFake	688	54	742

Table 1.39: Simulation results, where any difference in death rates between `treatmentFake` and `controlFake` is purely due to chance.

⊙ **Exercise 1.37** What is the difference in death rates between the two fake groups in Table 1.39? How does this compare to the observed 2.5% in the real groups? Answer in the footnote[30].

1.6.2 Checking for independence*

We computed one possible difference under the independence model in Exercise 1.37, which represents one difference due to chance. We could repeat the simulation to get another difference from chance: -0.005. And another: -0.010. And another: 0.003. And so on until we repeat the simulation enough times that we have a good idea of what represents the *distribution of differences from chance alone*. Figure 1.40 shows a histogram of the differences found from 100 simulations.

● **Example 1.9** How often would you observe a difference of 2.5% (0.025) according to Figure 1.40? Often, sometimes, rarely, or never?

It appears that a difference due to chance alone would only happen about 3% of the time according to Figure 1.40. We might describe that as being rare.

The difference of 2.5% is a rare event, and this suggests two possible interpretations of the results of the study:

H_0 **Independence model.** The drug doesn't work, and we observed a difference that would only happen rarely.

[29] These label counts correspond to the number of `treatment` and `control` assignments in the actual study.

[30] $54/742 - 47/733 = 0.0087$ or about 0.9%. This difference due to chance is smaller. However, we should run more simulations to get a good idea of what differences we get by chance, i.e. it is possible 0.9% was a fluke.

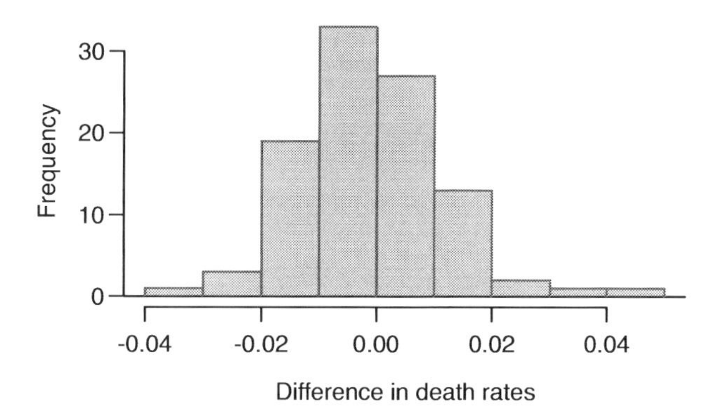

Figure 1.40: A histogram of differences from 100 simulations produced under the independence model, H_0, where `groupFake` and `outcome` are independent. Four of the one-hundred simulations had a difference of at least 2.5%.

H_A **Alternative model.** The drug does work, and what we observed was that the drug was actually working, which explains the large difference of 2.5%.

We often reject the notion that what we observe just happens to be rare, so in this case we would reject the independence model in favor of the alternative: we would conclude the drug works. That is, the data provides strong evidence that the drug actually is effective at reducing the death rate in heart attack patients.

One field of statistics is built on evaluating whether such differences are due to chance: statistical inference. In statistical inference, statisticians evaluate which model is most reasonable given the data. Errors do occur: we might choose the wrong model. While we do not always choose correctly, statistical inference gives us tools to control and evaluate how often these errors occur. In Chapter 4, we give a formal introduction to the problem of model selection. However, we spend the next two chapters building a foundation of probability and theory necessary to make that discussion rigorous.

1.7 Problem set

In Exercises 1 and 2 identify the cases and the variables studied as well as the main research question of the study.

1.1 Researchers collected data to examine the relationship between pollutants and preterm births in Southern California. Measurements of carbon monoxide (CO), nitrogen dioxide, ozone, and particulate matter less than 10μm (PM_{10}) at air-quality-monitoring stations were used to create exposure of estimates for periods of pregnancy. Birth weight data was also collected on 143,196 births between the years 1989 and 1993. The analysis suggested that increased ambient PM_{10} and, to a lesser degree, CO concentrations may contribute to the occurrence of preterm births. [1]

1.2 The Buteyko method is a shallow breathing technique developed by Konstantin Buteyko, a Russian doctor, in 1952. Anecdotal evidence suggests that the Buteyko method can reduce asthma symptoms and improve quality of life. In a study aimed to determine the effectiveness of this method, 600 adult patients aged 18-69 years diagnosed and currently treated for asthma were divided into two groups. One group practiced the Buteyko method and the other did not. Patients were scored on quality of life, activity, asthma symptoms and medication reduction. On average the participants in the Buteyko group experienced a significant improvement in asthma with a reduction in symptom and medication and improvement in quality of life. [2]

1.3 A survey was conducted to study the smoking habits of UK residents. Below is a data matrix displaying a portion of the data collected in this survey [3].

	gender	age	maritalStatus	grossIncome	smoke	amtWeekends	amtWeekdays
1	Female	42	Single	Under £2,600	Yes	12 cig/day	12 cig/day
2	Male	44	Single	£10,400 to £15,600	No	N/A	N/A
3	Male	53	Married	Above £36,400	Yes	6 cig/day	6 cig/day
⋮	⋮	⋮	⋮	⋮	⋮	⋮	⋮
1691	Male	40	Single	£2,600 to £5,200	Yes	8 cig/day	8 cig/day

(a) What does each row of the data matrix represent?

(b) How many participants were included in the survey?

(c) Identify the variables in the data set and give a short description of what each variable represents.

1.4 Exercise 1 describes a study of effects of air pollution exposure on preterm births in Southern California. Identify the population and the sample in this study.

1.5 Exercise 2 describes a study on the Buteyko method. Identify the population and the sample in this study.

1.6 A Statistics student curious about the relationship between amount of time students spend on social networking sites and their performance at school decides to conduct a survey. Indicate the sampling method for each of the techniques he is considering, as well as any bias you may expect.

(a) He randomly samples 40 students, gives them the survey, asks them to fill it out and bring it back the next day.

(b) He gives out the survey only to his friends, makes sure each one of them fills out the survey.

1.7 The Gallup Poll uses a procedure called random digit dialing (RDD) which creates phone numbers based on a list of all area codes in America, along with estimates of the number of residential households those exchanges have attached to them. This procedure is a lot more complicated than obtaining a list of all phone numbers from the phone book. Why does the Gallup Poll choose to use RDD instead of picking phone numbers from the phone book?

1.8 Exercise 3 introduced a study about the smoking habits of UK residents. Indicate if the variables in the study are numerical or categorical. If numerical, identify as continuous or discrete.

1.9 Sir Ronal Aylmer Fisher was an English statistician, evolutionary biologist and geneticist who worked on a data set that contained sepal length and width and petal length and width from three species of iris flowers (setosa, versicolor and virginica). There were 50 flowers from each species in the data set [4].

(a) How many cases were included data?

(b) How many numerical variables are included in the data? Indicate what they are and if they are continuous or discrete.

(c) How many numerical variables are included in the data and what are they? How many levels does each have, and what are they? Are they also ordinal?

Iris versicolor. Photocredits: rtclauss on Flickr
(http://www.flickr.com/photos/rtclauss/3834965043/

1.10 Data has been collected on life spans (in years) and gestation lengths (in days) for 62 mammals [5]. A scatter plot of life span vs. length of gestation is shown below.

(a) What type of an association is apparent between life span and length of gestation?

(b) What type of an association would you expect to see if the axes of the plot were reversed, i.e. if we plotted length of gestation vs. life span?

(c) Are life span and length of gestation independent? Explain your reasoning.

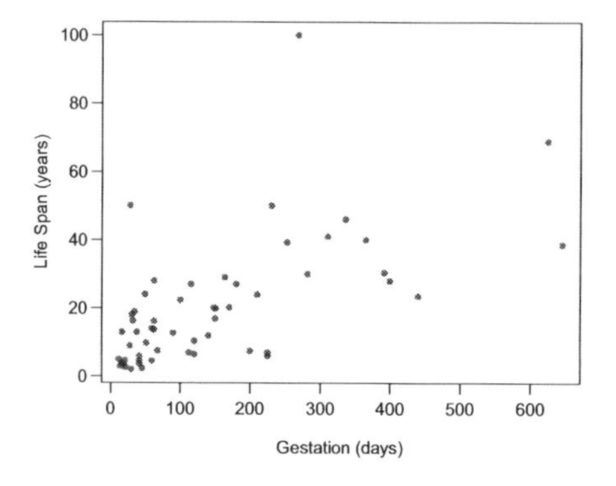

1.11 Office productivity is relatively low when the employees feel no stress about their work or job security. However high levels of stress can also lead to reduced employee productivity. Sketch a plot to represent the relationship between stress and productivity and explain your reasoning.

1.12 Indicate which of the below plots show a

(a) linear association

(b) non-linear association

(c) positive association

(d) negative association

(e) no association

Each part may refer to more than one plot.

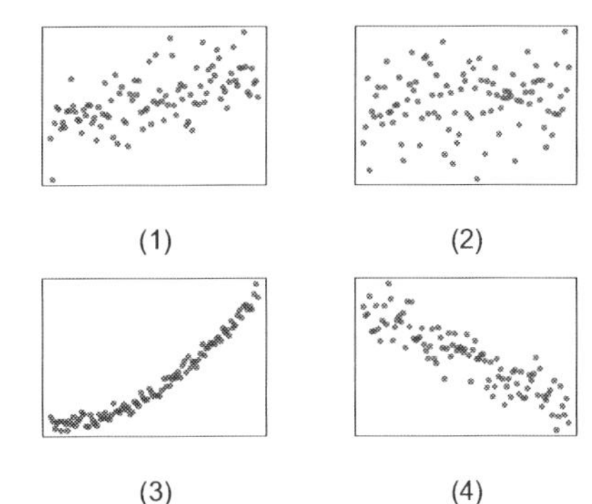

1.13 Identify which value represents the sample mean and which value represents the claimed population mean.

(a) A recent article in a college newspaper stated that college students get an average of 5.5 hrs of sleep each night. A student who was skeptical about this value decided to conduct a survey on a random sample of 25 students which yielded an average of 6.25 hrs of sleep.

(b) American households spent an average of about $52 in 2007 on Halloween merchandise such as costumes, decorations and candy. To see if this number has changed, researchers conducted a new survey in 2008 on 1,500 households and found that in the average amount a household spent on Halloween was $58.

1.14 A random sample from the data set introduced in Exercise 3 is given. Find the mean for each of the variables listed and mark it on the dot plot of the variable. If you cannot calculate the mean for a variable, indicate why.

gender	age	maritalStatus	grossIncome	smoke	amtWeekends	amtWeekdays
Female	51	Married	2,600 to 5,200	Yes	20	20
Male	24	Single	10,400 to 15,600	Yes	20	15
Female	33	Married	10,400 to 15,600	Yes	20	10
Female	17	Single	5,200 to 10,400	Yes	20	15
Female	76	Widowed	5,200 to 10,400	Yes	20	20

1.15 Exercise 3 introduces a data set on the smoking habits of UK residents. Below are histograms displaying the distributions of amount of cigarettes smoked on weekdays and on weekends. Describe the two distributions and compare them.

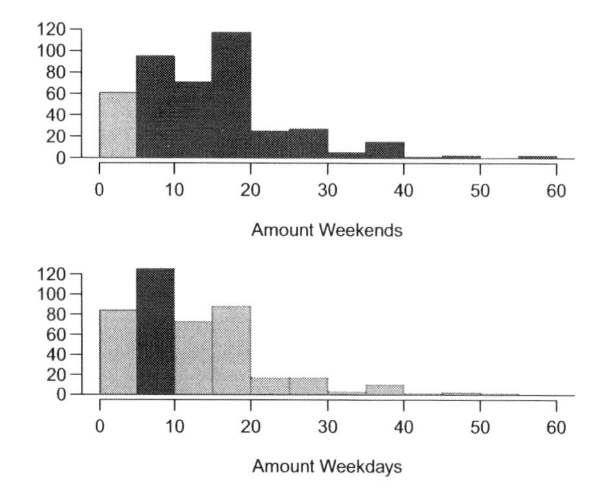

1.16 Below are the final scores of 20 introductory statistics students.

79, 83, 57, 82, 94, 83, 72, 74, 73, 71,
66, 89, 78, 81, 78, 81, 88, 69, 77, 79

Draw a histogram of these data and describe the distribution.

1.17 Find the standard deviation of the amount of cigarettes smoked on weekdays and on weekends by the 5 randomly sampled UK residents given in Exercise 14. Is the variability higher on weekends or on weekdays?

1.18 Find the median in following data sets.

(a) 3, 5, 6, 7, 9

(b) 3, 5, 6, 7, 9, 10

1.19 Below is the five number summary for the final exam scores data given in Exercise 16. Create a box plot of the data based on these values.

Min	57
Q1	72.5
Q2 (Median)	78.5
Q3	82.5
Max	94

1.20 Describe the distribution in the following histograms and match them to the box plots.

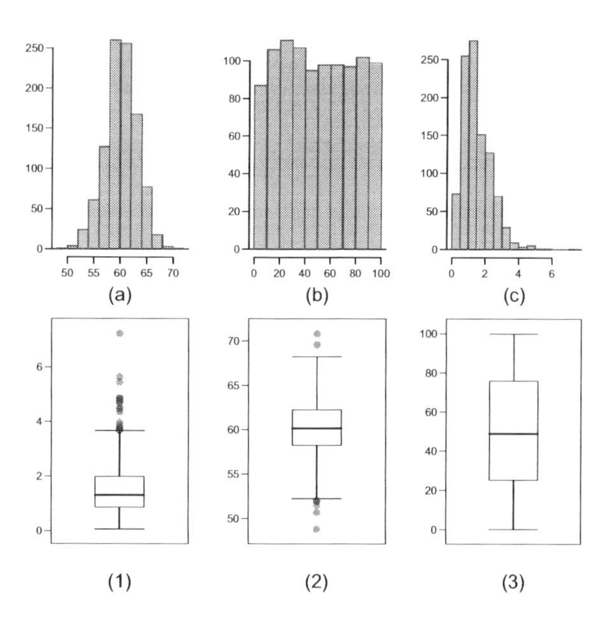

1.21 Compare the two plots given below. What characteristics of the distribution are apparent in the histogram and not in the box plot? What characteristics are apparent in the box plot but not in the histogram?

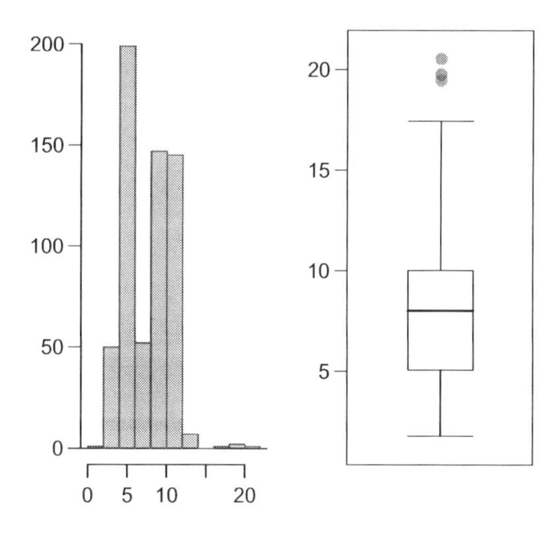

1.22 Below are a histogram and box plot of marathon times of male and female New York Marathon runners between 1980 and 1999.

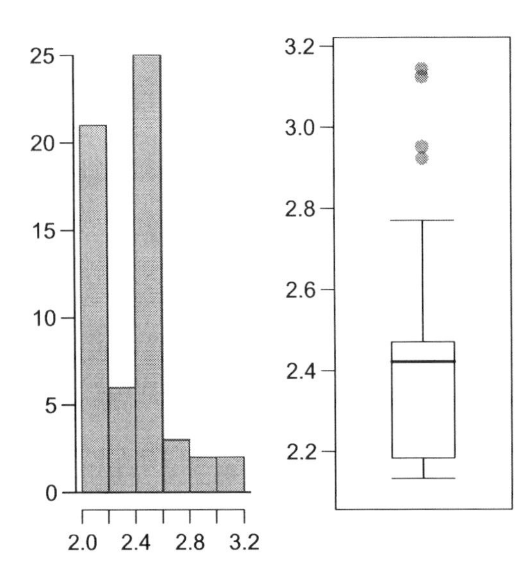

(a) What features of the distribution are apparent in the histogram and not the box plot? What features are apparent in the box plot but not in the histogram?

(b) What may be the reason for the bimodal distribution? Explain.

(c) Compare the distribution of marathon times for men and women based on the box plots given below.

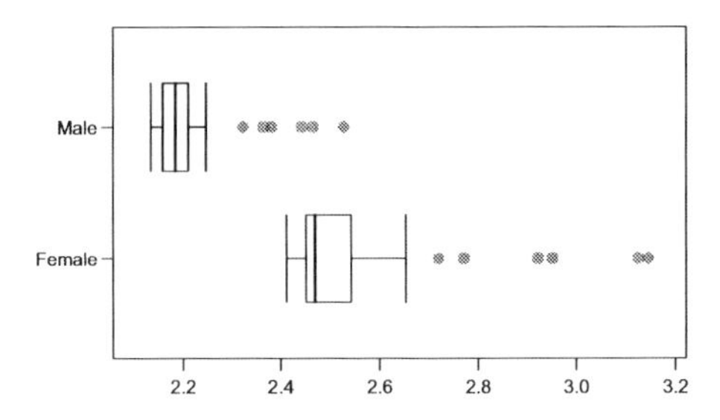

1.23 Take another look at the data presented in Exercise 22 using the time series plot below. Describe the trends apparent in the plot and comment on whether or not these trends were apparent in the histogram and/or the box plot.

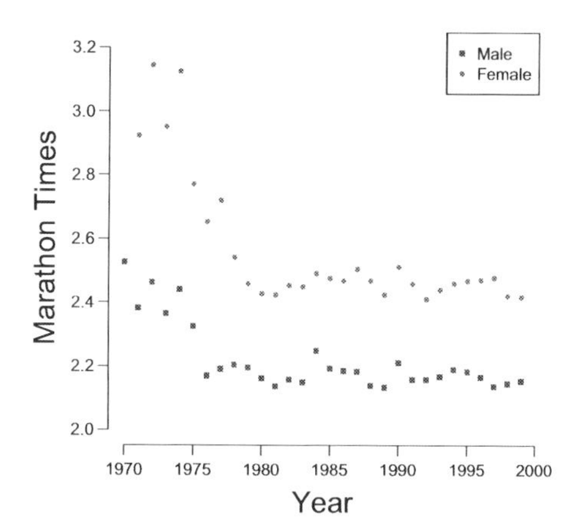

1.24 Describe whether you expect the distribution to be symmetric, right skewed or left skewed. In each case, specify whether you would use the mean or median to describe the center, and whether you would prefer to use the standard deviation or IQR to describe the spread.

(a) Housing prices in a country where 25% of the houses cost below \$350,000, 50% of the houses cost below \$450,000, 75% of the houses cost below \$1,000,000 and there are houses selling at over \$6,000,000.

(b) Housing prices in a country where 25% of the houses cost below \$300,000, 50% of the houses cost below \$600,000, 75% of the houses cost below \$900,000 and there no houses selling at over \$1,200,000.

(c) Number of alcoholic drinks consumed by college students in a given week.

(d) Annual salaries of the employees of a Fortune 500 company.

1.25 The first histogram below shows the distribution of the yearly incomes of 40 patrons at a coffee shop. Then walks in two friends, one makes \$225,000 per year and the other makes \$250,000. The second histogram shows the new distribution of income. Also provided are summary statistics for the two distributions.

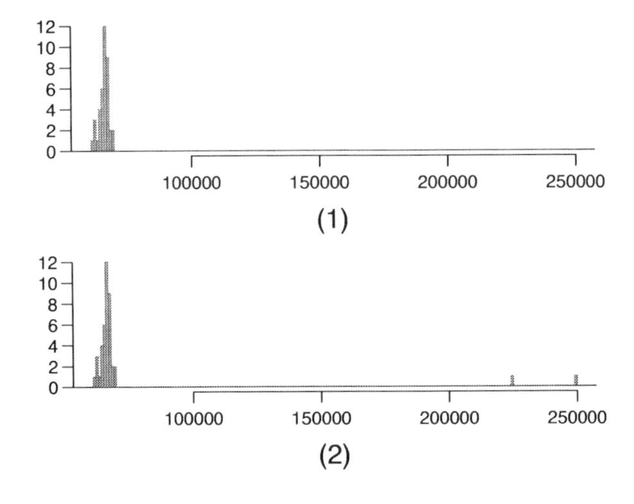

	(1)	(2)
n	40	42
Min.	60,420	60,420
1st Qu.	64,450	64,610
Median	65,210	65,370
Mean	65,060	73,270
3rd Qu.	66,290	66,320
Max.	68,320	250,000
SD	1,832	37,313

(a) Is the mean or the median a better measure of the typical amount earned by the 42 patrons at this coffee shop? What does this say about the robustness of the two measures.

(b) Is the standard deviation or the IQR a better measure of variability in the amounts earned by the 42 patrons at this coffee shop? What does this say about the robustness of the two measures.

1.26 Below is a relative frequency bar plot and a pie chart showing the distribution of marital status in the data set on the smoking habits of UK residents introduced in Exercise 3.

(a) What features are apparent in the bar plot but not in the pie chart?

(b) What features are apparent in the pie chart but not in the bar plot?

(c) Which graph would you prefer to use for displaying categorical data? Consider which graph gives you the most information.

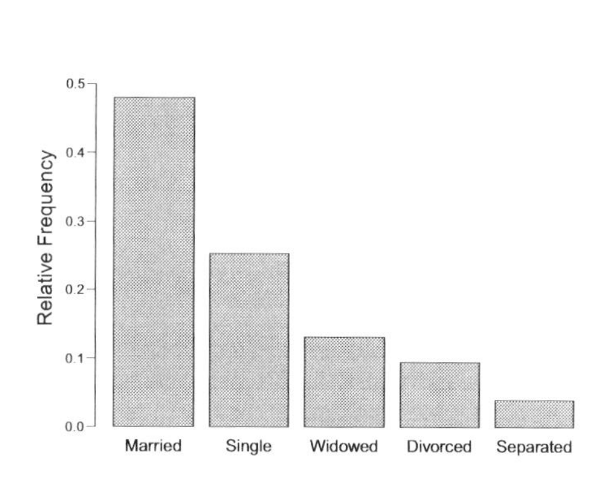

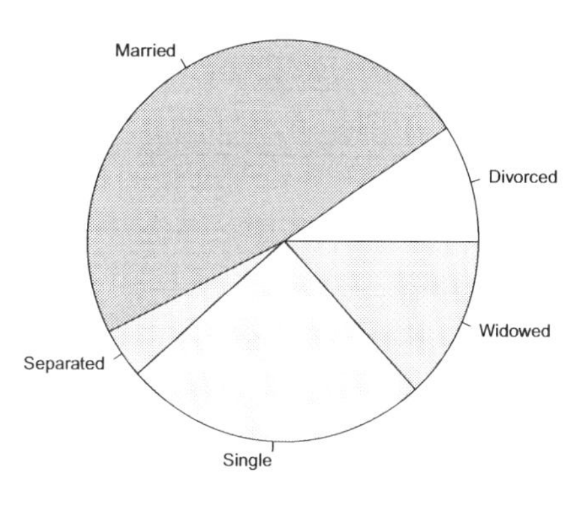

1.27 The table below shows the relationship between hair color and eye color for a group of 1,770 German men.

		Hair Color			
		Brown	Black	Red	Total
Eye	Brown	400	300	20	720
Color	Blue	800	200	50	1050
	Total	1200	500	70	1770

(a) What percentage of the men have black hair?

(b) What percentage of the men have blue eyes?

(c) What percentage of the men with black hair have blue eyes?

(d) Does it appear hair and eye color are independent? If not, how are they associated?

1.28 Exercise 3 introduces a data set on the smoking habits of UK residents.

		Gender	
		Female	Male
Smoke	No	731	539
	Yes	234	187

(a) What percent of women smoke?

(b) What percent of men smoke?

(c) Based on these percentages, is smoking independent of gender?

1.29 Exercise 3 introduces a data set on the smoking habits of UK residents. Based on the mosaic plot below, is smoking independent of marital status?

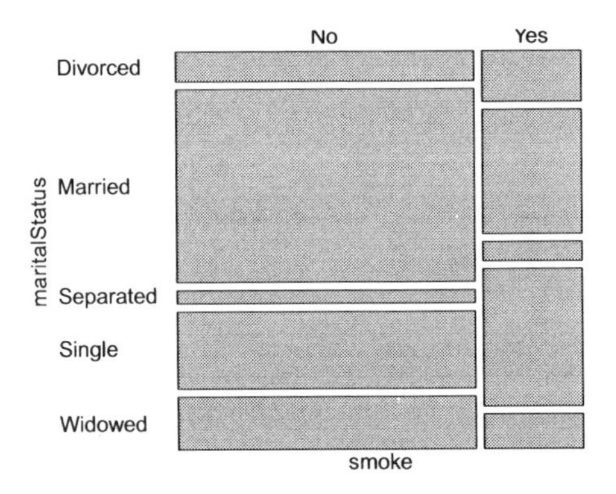

1.30 The Stanford University Heart Transplant Study was conducted to determine whether an experimental heart transplant program increased lifespan [6]. Each patient entering the program was designated officially a heart transplant candidate, meaning that he was gravely ill and would most likely benefit from a new heart. Some patients got a transplant and some did not. The variable `transplant` indicates what group the patients were in; treatment group got a transplant and control group did not. Another variable in the study, `survived`, indicates whether or not the patient was alive at the end of the study. Based on the mosaic plot below, is survival independent of whether or not the patient got a transplant? Explain your reasoning.

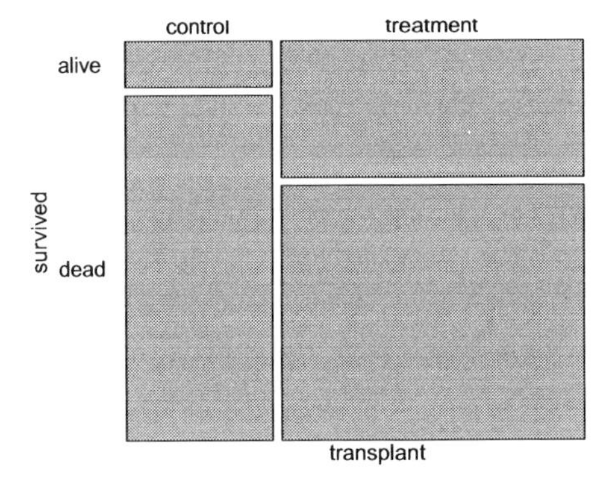

1.31 Exercise 30 introduces the Stanford Heart Transplant Study. Below are side-by-side box plots of the survival times (in days) of patients in the control and treatment groups. Write a few sentences about what you see. What can we conclude from the results of this study?

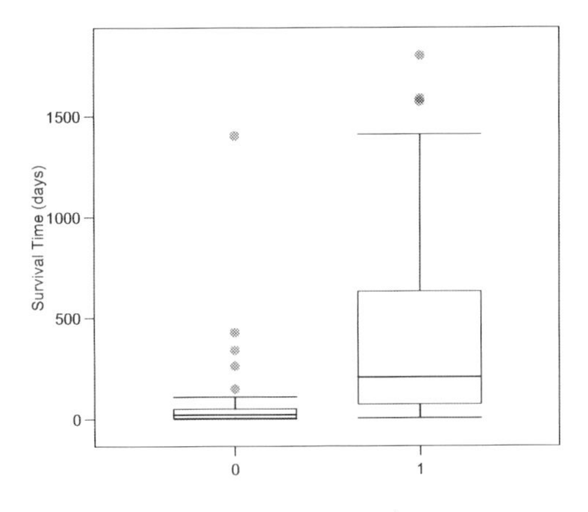

1.32 Below is a scatter plot displaying the relationship between the number of hours per week students watch TV and the grade they got in a statistics class (out of 100).

(a) What is the explanatory variable and what is the response variable?

(b) Is this an experiment or an observational study?

(c) Describe the relationship between the two variables.

(d) Can we conclude that watching longer hours of TV causes the students to get lower grades?

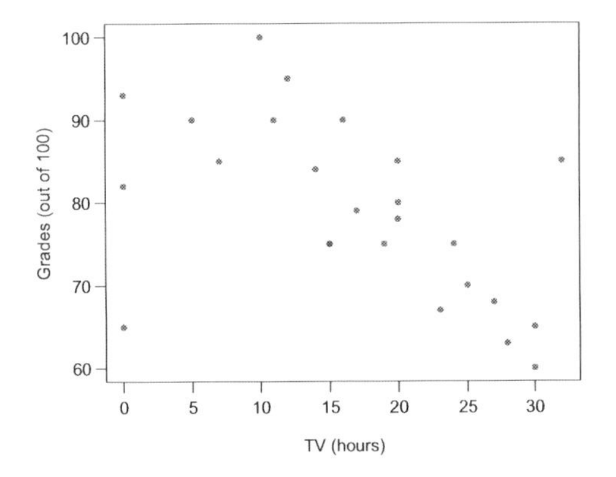

1.33 In order to assess the effectiveness of a vitamin supplements, researchers prescribed a certain vitamin to patients. After 6 months, the researchers asked the patients whether or not they have been taking the vitamin. Then they divided the patients into two groups, those who took the pills and those who did not and compared some health conditions between the two groups to measure the effectiveness of the vitamin.

(a) Was this an experiment or an observational study? Why?

(b) What are the explanatory and response variables in this study?

(c) This study is seriously flawed. Use the language of statistics to explain the flaw and how this affects the validity of the conclusion reached by the researchers.

(d) Were the patients blinded to their treatment? If not, explain how to make this a blinded study.

(e) Were the researchers blinded in this study? If not, explain how to make this a double blind study.

1.34 Exercise 30 introduces the Stanford Heart Transplant Study. Of the 34 patients in the control group 4 were alive at the end of the study and of the 69 patients in the treatment group 24 were alive. A researcher wrote whether or not each patient survived on 3×5 cards, shuffled all the cards together, then dealt them into two groups of size 69 and 34. After shuffling, there were 9 patients in the control group and 19 patients in the treatment group who survived. The below tables summarize the results. What does this tell us about the success of the heart transplant program?

Data		
	control	treatment
alive	4	24
dead	30	45

Independence Model		
	control	treatment
alive	9	19
dead	25	50

1.35 The researcher in Exercise 34 then simulates chance models 250 times and each time records the number of people in the treatment group who died. Below is a histogram of these counts.

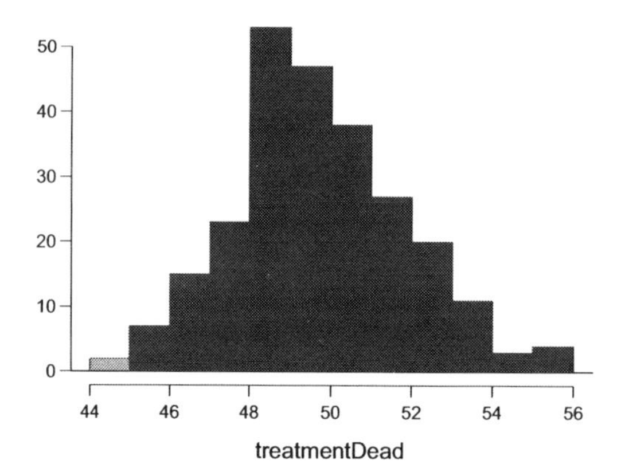

(a) What are the hypotheses?

(b) What does this say about the effectiveness of the transplant program.

Chapter 2

Probability

Probability forms a foundation for statistics. You may already be familiar with many aspects of probability, however, formalization of the concepts is new for most. This chapter aims to introduce probability on familiar terms using processes most people have encountered.

2.1 Defining probability

● **Example 2.1** A "die", the singular of dice, is a cube with six faces numbered `1`, `2`, `3`, `4`, `5`, and `6`. What is the chance of getting `1` when rolling a die?

If the die is fair, then the chance of a `1` is as good as the chance of any other number. Since there are six outcomes, the chance must be 1-in-6 or, equivalently, $1/6$.

● **Example 2.2** What is the chance of getting a `1` or `2` in the next roll?

`1` and `2` constitute two of the six equally likely possible outcomes, so the chance of getting one of these two outcomes must be $2/6 = 1/3$.

● **Example 2.3** What is the chance of getting either `1`, `2`, `3`, `4`, `5`, or `6` on the next roll?

100%. The die must be one of these numbers.

● **Example 2.4** What is the chance of not rolling a `2`?

Since the chance of rolling a `2` is $1/6$ or $16.\bar{6}\%$, the chance of not getting a `2` must be $100\% - 16.\bar{6}\% = 83.\bar{3}\%$ or $5/6$.

Alternatively, we could have noticed that not rolling a `2` is the same as getting a `1`, `3`, `4`, `5`, or `6`, which makes up five of the six equally likely outcomes and has probability $5/6$.

● **Example 2.5** Consider rolling two dice. If $1/6^{th}$ of the time the first die is `1` and $1/6^{th}$ of those times the second die is a `1`, what is the chance of getting two `1`s?

If $16.\bar{6}\%$ of the time the first die is a `1` and $1/6^{th}$ of *those* times the second die is also a `1`, then the chance both dice are `1` is $(1/6) * (1/6)$ or $1/36$.

2.1.1 Probability

We use probability to build tools to describe and understand apparent randomness. We often frame probability in terms of a **random process** giving rise to an **outcome**.

Roll a die	$\rightarrow$	`1`, `2`, `3`, `4`, `5`, or `6`
Flip a coin	$\rightarrow$	`H` or `T`

Rolling a die or flipping a coin is a seemingly random process and each gives rise to an outcome.

> **Probability**
> The **probability** of an outcome is the proportion of times the outcome would occur if we observed the random process an infinite number of times.

Probability is defined as a proportion, and it must always be between 0 and 1 (inclusively). It may also be displayed as a percentage between 0% and 100%.

Probability can be illustrated by rolling a die many times. Let $\hat{p}_n$ be the proportion of outcomes that are `1` after the first n rolls. As the number of rolls increases, $\hat{p}_n$ will converge to the probability of rolling `1`, $p = 1/6$. Figure 2.1 shows this convergence for 100,000 die rolls. The tendency of $\hat{p}_n$ to stabilize around p is described by the **Law of Large Numbers**.

> **Law of Large Numbers**
> As more observations are collected, the proportion $\hat{p}_n$ of occurrences with a particular outcome converges to the probability p of that outcome.

Occasionally the proportion will veer off from the probability and appear to defy the Law of Large Numbers, as $\hat{p}_n$ does many times in Figure 2.1. However, these deviations become smaller as the sample size becomes larger.

Above we write p as the probability of rolling a `1`. We can also write this probability as

$$P(\text{rolling a } 1)$$

As we become more comfortable with this notation, we will abbreviate it further. For instance, if it is clear the process is "rolling a die", we could abbreviate P(rolling a `1`) as $P(1)$.

⊙ **Exercise 2.1** Random processes include rolling a die and flipping a coin. (a) Think of another random process. (b) Describe all the possible outcomes of that process.

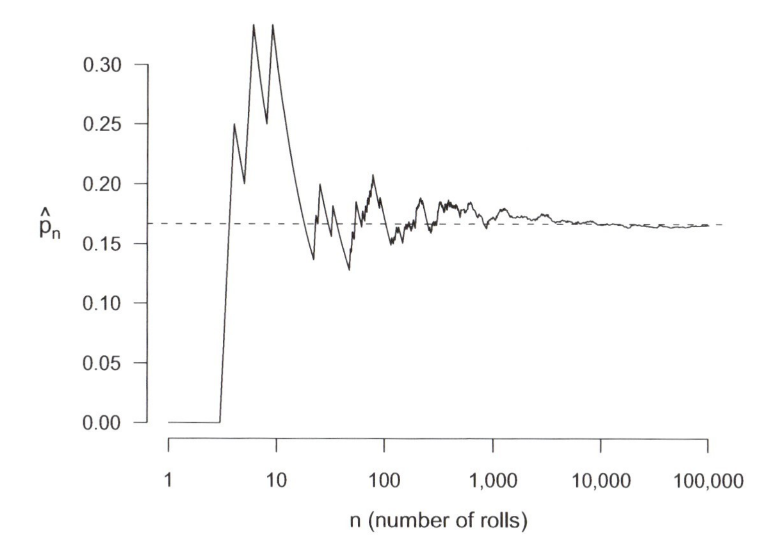

Figure 2.1: The fraction of die rolls that are 1 at each stage in a simulation. The proportion tends to get closer to the probability $1/6 \approx 0.167$ as the sample size gets large.

For instance, rolling a die is a random process with potential outcomes `1`, `2`, ..., `6`. Several examples are in the footnote[1].

What we think of as random processes are not necessarily random but may only be too difficult to understand. Example (iv) in Exercise 2.1 suggests whether your roommate does her dishes tonight is random. However, even if your roommate's behavior is not truly random, modeling her behavior as a random process can still be useful.

TIP: Modeling a process as random
It can be helpful to model a process as random even if it is not truly random.

2.1.2 Disjoint or mutually exclusive outcomes

Two outcomes are called **disjoint** or **mutually exclusive** if they cannot both happen. For instance, if we roll a die, the outcomes `1` and `2` are disjoint since they cannot both occur. On the other hand, the outcomes `1` and "rolling an odd

[1](i) Whether someone gets sick in the next month or not is an apparently random process with outcomes `sick` and `not`. (ii) We can *generate* a random process by randomly picking a person and measuring that person's height. The outcome of this process will be a positive number. (iii) Whether the stock market goes up or down next week is a seemingly random process with possible outcomes `up`, `down`, and `noChange`. (iv) Whether your roommate cleans her dishes tonight probably seems like a random process with possible outcomes `cleansDishes` and `leavesDishes`.

number" are not disjoint since both occur if the die is a 1. The terms *disjoint* and *mutually exclusive* are equivalent and interchangeable.

Calculating the probability of disjoint outcomes is easy. When rolling a die, the outcomes 1 and 2 are disjoint, and we compute the probability that one of these outcomes will occur by adding their separate probabilities:

$$P(1 \text{ or } 2) = P(1) + P(2) = 1/6 + 1/6 = 1/3$$

What about the probability of rolling a 1, 2, 3, 4, 5, or 6? Here again, all of the outcomes are disjoint so we add the probabilities:

$$\begin{aligned} &P(1 \text{ or } 2 \text{ or } 3 \text{ or } 4 \text{ or } 5 \text{ or } 6) \\ &\quad = P(1) + P(2) + P(3) + P(4) + P(5) + P(6) \\ &\quad = 1/6 + 1/6 + 1/6 + 1/6 + 1/6 + 1/6 = 1. \end{aligned}$$

The **Addition Rule** guarantees the accuracy of this approach when the outcomes are disjoint.

Addition Rule of disjoint outcomes

If OC_1 and OC_2 represent two disjoint outcomes, then the probability that one of them occurs is given by

$$P(OC_1 \text{ or } OC_2) = P(OC_1) + P(OC_2)$$

If there are many disjoint outcomes OC_1, ..., OC_k, then the probability of one of these outcomes occurring is

$$P(OC_1) + P(OC_2) + \cdots + P(OC_k) \qquad (2.1)$$

⊙ **Exercise 2.2** We are interested in the probability of rolling a 1, 4, or 5. (a) Explain why the outcomes 1, 4, and 5 are disjoint. (b) Apply the Addition Rule for disjoint outcomes to determine $P(1 \text{ or } 4 \text{ or } 5)$.

⊙ **Exercise 2.3** In the `cars` data set in Chapter 1, the `type` variable described the size of the vehicle: `small` (21 cars), `midsize` (22 cars), or `large` (11 cars). (a) Are the outcomes `small`, `midsize`, and `large` disjoint? Answer in the footnote[2]. (b) Determine the proportion of `midsize` and `large` cars separately. (c) Use the Addition Rule for disjoint outcomes to compute the probability a randomly selected car from this sample is either `midsize` or `large`.

Statisticians rarely work with individual outcomes and instead consider *sets* or *collections* of outcomes. Let A represent the event where a die roll results in 1 or 2 and B represent the event that the die roll is a 4 or a 6. We write A as the

[2]Yes. Each car is categorized in only one level of `type`.

set of outcomes {1, 2} and $B = \{4, 6\}$. These sets are commonly called **events**. Because A and B have no elements in common, they are disjoint events. A and B are represented in Figure 2.2.

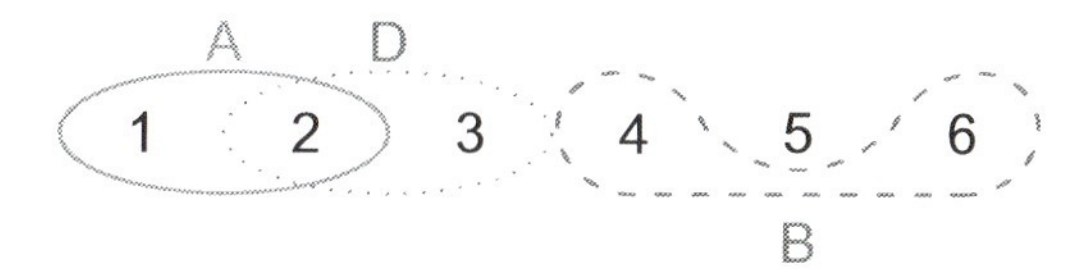

Figure 2.2: Three events, A, B, and D, consist of outcomes from rolling a die. A and B are disjoint since they do not have any outcomes in common. What other pair of events are disjoint?

The Addition Rule applies to both disjoint outcomes and disjoint events. The probability that one of the disjoint events A or B occurs is the sum of the separate probabilities:

$$P(A \text{ or } B) = P(A) + P(B) = 1/3 + 1/3 = 2/3$$

⊙ **Exercise 2.4** (a) Verify the probability of event A, $P(A)$, is $1/3$ using the Addition Rule for disjoint events. (b) Do the same for B.

⊙ **Exercise 2.5** (a) Using Figure 2.2 as a reference, what outcomes are represented by the event D? (b) Are events B and D disjoint? (c) Are events A and D disjoint? Answer to part (c) is in the footnote[3].

⊙ **Exercise 2.6** In Exercise 2.5, you confirmed B and D from Figure 2.2 are disjoint. Compute the probability that either event B or event D occurs.

Let's consider calculations for two events that are not disjoint in the context of a regular deck of 52 cards, represented in Table 2.3. If you are unfamiliar with the cards in a regular deck, please see the footnote[4].

⊙ **Exercise 2.7** (a) What is the probability a random card is a diamond? (b) What is the probability a randomly selected card is a face card?

Venn diagrams are useful when outcomes can be categorized as "in" or "out" for two or three variables, attributes, or random processes. The Venn diagram in Figure 2.4 uses a circle to represent diamonds and another to represent face cards. If a card is both a diamond and a face card, it falls into the intersection of the

[3] The events A and D share an outcome in common, 2, and so are not disjoint.

[4] The 52 cards are split into four **suits**: ♣ (club), ♢ (diamond), ♡ (heart), ♠ (spade). Each suit has its 13 cards labeled: `2`, `3`, ..., `10`, `J` (jack), `Q` (queen), `K` (king), and `A` (ace). Thus, each card is a unique combination of a suit and a label, e.g. 4♡. The 12 cards represented by the jacks, queens, and kings are called `face cards`. The cards that are ♢ or ♡ are typically colored `red` while the other two suits are typically colored `black`.

2♣	3♣	4♣	5♣	6♣	7♣	8♣	9♣	10♣	J♣	Q♣	K♣	A♣
2♢	3♢	4♢	5♢	6♢	7♢	8♢	9♢	10♢	J♢	Q♢	K♢	A♢
2♡	3♡	4♡	5♡	6♡	7♡	8♡	9♡	10♡	J♡	Q♡	K♡	A♡
2♠	3♠	4♠	5♠	6♠	7♠	8♠	9♠	10♠	J♠	Q♠	K♠	A♠

Table 2.3: Representations of the 52 unique cards in a deck.

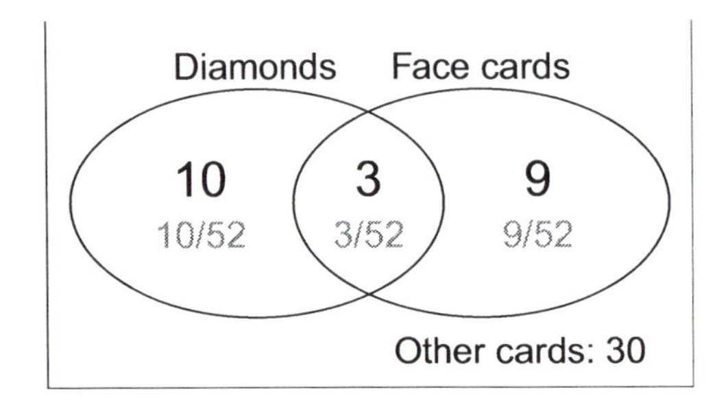

Figure 2.4: A Venn diagram for diamonds and face cards.

circles. If it is a diamond but not a face card, it will be in the left circle but not in the right (and so on). The total number of cards that are diamonds is given by the total number of cards in the diamonds circle: $13 + 3 = 13$. The probabilities are shown in gray.

⊙ **Exercise 2.8** Using the Venn diagram, verify $P(\text{face card}) = 12/52 = 3/13$.

Let A represent the event a randomly selected card is a diamond and B represent the event it is a face card. How do we compute $P(A \text{ or } B)$? Events A and B are not disjoint – the cards $J\diamondsuit$, $Q\diamondsuit$, and $K\diamondsuit$ fall into both categories – so we cannot use the Addition Rule for disjoint events. Instead we use the Venn diagram. We start by adding the probabilities of the two events:

$$P(A) + P(B) = P(\diamondsuit) + P(\text{face card}) = 12/52 + 13/52$$

However, the three cards that are in both events were counted twice, once in each probability. We correct this double counting using a correction term:

$$\begin{aligned} P(A \text{ or } B) &= P(\text{face card or } \diamondsuit) \\ &= P(\text{face card}) + P(\diamondsuit) - P(\text{face card } \& \diamondsuit) \qquad (2.2) \\ &= 12/52 + 13/52 - 3/52 \\ &= 22/52 = 11/26 \end{aligned}$$

Equation (2.2) is an example of the **General Addition Rule**.

General Addition Rule
If A and B are any two events, disjoint or not, then the probability that at least one of them will occur is

$$P(A) + P(B) - P(A \& B) \tag{2.3}$$

where $P(A \& B)$ is the probability that both events occur.

TIP: "or" is inclusive
When we write *or* in statistics, we mean *and/or* unless we explicitly state otherwise. Thus, A or B occurs means A, B, or both A and B occur.

⊙ **Exercise 2.9** (a) If A and B are disjoint, then $P(A \& B) = 0$. Why? (b) Using part (a), verify that the General Addition Rule simplifies to the Addition Rule for disjoint events if A and B are disjoint. Answers in the footnote[5].

⊙ **Exercise 2.10** In the `cars` data set with 54 vehicles, 22 were `midsize` cars and 16 had a capacity of `6` people. However, 5 `midsize` cars had a capacity of `6` people. Create a Venn diagram for this setup. [6].

⊙ **Exercise 2.11** (a) Use your Venn diagram from Exercise 2.10 to determine the probability a random car from the `cars` data set is both a `midsize` vehicle and has a capacity of `6`. (b) What is the probability the car is a `midsize` car or has a `6` person capacity?

2.1.3 Probability distributions

The grocery receipt for a college student is shown in Table 2.5. Does anything seem odd about the total? The individual costs only add up to \$23.20 while the total is \$37.90. Where did the additional \$14.70 come from?

Table 2.6 shows another month of expenditures with a new problem. While the sum of the expenditures match up, the amount spent on milk is a negative amount!

A **probability distribution** is a table of all disjoint outcomes and their associated probabilities. It is like a grocery bill, except that instead of foods there are

[5](a) If A and B are disjoint, A and B can never occur simultaneously. (b) If A and B are disjoint, then the last term of Equation (2.3) is 0 (see part (a)) and we are left with the Addition Rule for disjoint events.

[6]

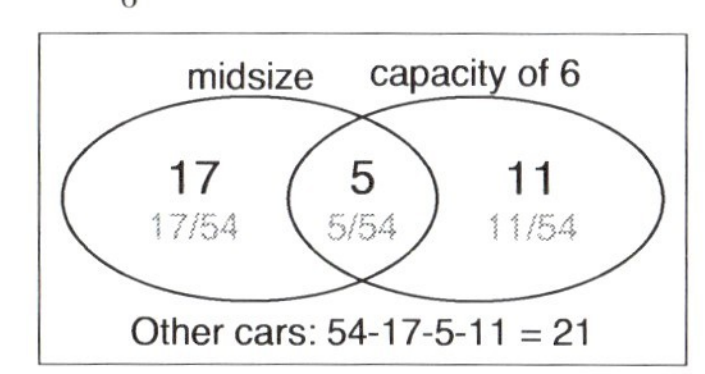

Item	Cost
Spaghetti	$6.50
Carrots	$7.20
Apples	$3.10
Milk	$6.40
Tax	$0.00
Total	$37.90

Table 2.5: Grocery receipt with a total greater than the sum of the costs.

Item	Cost
Spaghetti	$6.50
Carrots	$7.20
Apples	$5.10
Milk	-$4.40
Chocolate	$1.19
Tax	$0.11
Total	$15.70

Table 2.6: On this receipt, buying milk saves the customer money (!).

outcomes, and instead of costs for each food item there are probabilities for each outcome. Table 2.7 shows the probability distribution for the sum of two dice.

Dice sum	2	3	4	5	6	7	8	9	10	11	12
Probability	$\frac{1}{36}$	$\frac{2}{36}$	$\frac{3}{36}$	$\frac{4}{36}$	$\frac{5}{36}$	$\frac{6}{36}$	$\frac{5}{36}$	$\frac{4}{36}$	$\frac{3}{36}$	$\frac{2}{36}$	$\frac{1}{36}$

Table 2.7: Probability distribution for the sum of two dice.

Probability distributions share the same structure as grocery receipts. However, probability distributions impose one special rule on the total: it must be 1.

Rules for probability distributions
A probability distribution is a list of the possible outcomes with corresponding probabilities that satisfies three rules:

1. The outcomes listed must be disjoint.
2. Each probability must be between 0 and 1.
3. The probabilities must total 1.

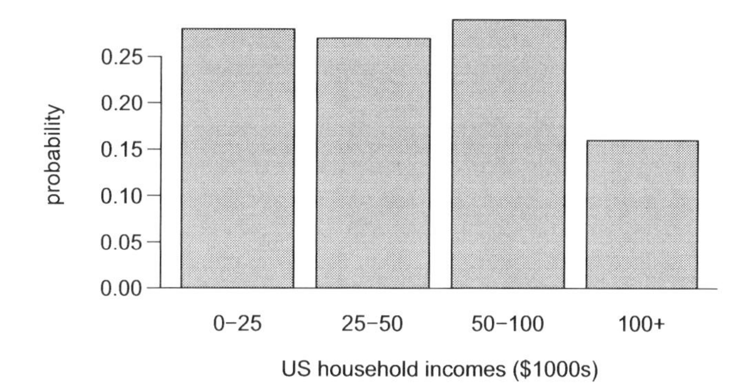

Figure 2.9: The probability distribution of US household income.

⊙ **Exercise 2.12** Table 2.8 suggests three distributions for household income in the United States. Only one is correct. Which one must it be? What is wrong with the other two? Answer in the footnote[7].

Income range ($1000s)	0-25	25-50	50-100	100+
(a)	0.18	0.39	0.33	0.16
(b)	0.38	-0.27	0.52	0.37
(c)	0.28	0.27	0.29	0.16

Table 2.8: Proposed distributions of US household incomes (Exercise 2.12).

Chapter 1 emphasized the importance of plotting data to provide quick summaries. Probability distributions can also be plotted in the form of a bar plot. For instance, the distribution of US household incomes is shown in Figure 2.9 as a bar plot[8]. The probability distribution for the sum of two dice is shown in Table 2.7 and plotted in Figure 2.10.

In these bar plots, the bar heights represent the outcome probabilities. If the outcomes are numerical and discrete, it is usually (visually) convenient to place the bars at their associated locations on the axis, as is the case of the sum of two dice. Another example of plotting the bars at their respective locations is shown in Figure 2.25 on page 81.

[7]The probabilities of (a) do not sum to 1. The second probability in (b) is negative. This leaves (c), which sure enough satisfies the requirements of a distribution. One of the three was said to be the actual distribution of US household incomes, so it must be (c).

[8]It is also possible to construct a distribution plot when income is not artificially binned into four groups. *Continuous* distributions are considered in Section 2.2.

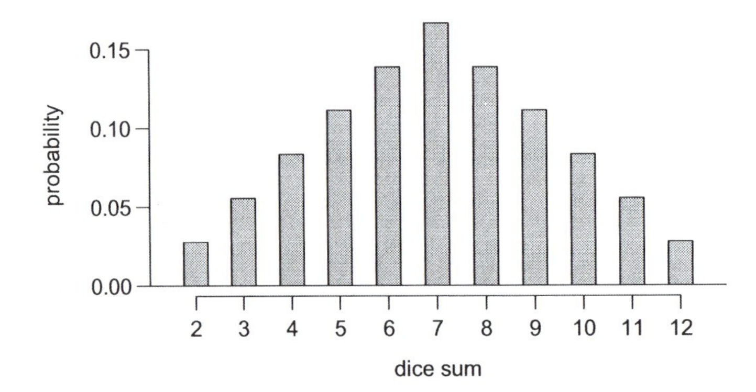

Figure 2.10: The probability distribution of the sum of two dice.

2.1.4 Complement of an event

Every simple process has an outcome. Rolling a die produces a value in the set {1, 2, 3, 4, 5, 6}. This set of all possible outcomes is called the **sample space** (S). We use the sample space to examine the scenario where an event does not occur.

Let $D = \{2, 3\}$ represent the event we roll a die and get 2 or 3. Then the **complement** of D represents all outcomes in our sample space that are not in D, which is denoted by $D^c = \{1, 4, 5, 6\}$. In other words, D^c is the set of all possible outcomes not already included in D. Figure 2.11 shows the relationship between D, D^c, and the sample space S.

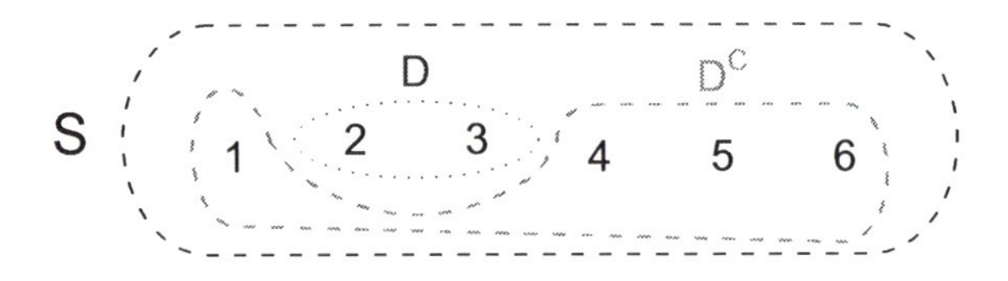

Figure 2.11: Event $D = \{2, 3\}$ and its complement, $D^c = \{1, 4, 5, 6\}$. S represents the sample space, which is the set of all possible events.

⊙ **Exercise 2.13** (a) Compute $P(D^c) = P(\text{rolling a 1, 4, 5, or 6})$. (b) What is $P(D) + P(D^c)$?

⊙ **Exercise 2.14** Events $A = \{1, 2\}$ and $B = \{4, 6\}$ are shown in Figure 2.2 on page 57. (a) Write out what A^c and B^c represent. (b) Compute $P(A^c)$ and $P(B^c)$. (c) Compute $P(A) + P(A^c)$ and $P(B) + P(B^c)$.

A complement of an event A is constructed to have two very important properties: (i) every possible outcome not in A is in A^c, and (ii) A and A^c are disjoint. Property (i) implies

$$P(A \text{ or } A^c) = 1 \tag{2.4}$$

That is, if the outcome is not in A, it must be represented in A^c. We use the Addition Rule for disjoint events to apply Property (ii):

$$P(A \text{ or } A^c) = P(A) + P(A^c) \tag{2.5}$$

Combining Equations (2.4) and (2.5) yields a very useful relationship between the probability of an event and its complement.

Complement

The complement of event A is denoted A^c, and A^c represents all outcomes not in A. A and A^c are mathematically related:

$$P(A) + P(A^c) = 1, \quad \text{i.e.} \quad P(A) = 1 - P(A^c) \tag{2.6}$$

In simple examples, computing A or A^c is feasible in a few simple steps. However, using the complement can save a lot of time as problems grow in complexity.

⊙ **Exercise 2.15** Let A represent the event where we roll two dice and their total is less than 12. (a) What does the event A^c represent? Answer in the footnote[9]. (b) Determine $P(A^c)$ from Table 2.7 on page 60. (c) Determine $P(A)$. Answer in the footnote[10].

⊙ **Exercise 2.16** Consider again the probabilities from Table 2.7 and rolling two dice. Find the following probabilities: (a) The sum of the dice is *not* 6. Hint in the footnote[11]. (b) The sum is at least 4. That is, determine the probability of the event $B = \{4, 5, ..., 12\}$. Answer in the footnote[12]. (c) The sum is no more than 10. That is, determine the probability of the event $D = \{2, 3, ..., 10\}$.

2.1.5 Independence

Just as variables and observations can be independent, random processes can be independent. Two processes are **independent** if knowing the outcome of one provides no useful information about the outcome of the other. For instance, flipping a coin and rolling a die are two independent processes – knowing the coin was H does not help determine what the die will be. On the other hand, stock prices usually move up or down together and so are not independent.

Example 2.5 provides a basic example of two independent processes: rolling two dice. We want to determine the probability that both will be 1. Suppose one of the dice is red and the other white. If the outcome of the red die is a

[9] The complement of A: when the total is equal to 12.

[10] Use the probability of the complement from part (b), $P(A^c) = 1/36$, and Equation (2.6): $P(\text{less than } 12) = 1 - P(12) = 1 - 1/36 = 35/36$.

[11] First find $P(6)$.

[12] We first find the complement, which requires much less effort: $P(2 \text{ or } 3) = 1/36+2/36 = 1/12$. Then we find $P(B) = 1 - P(B^c) = 1 - 1/12 = 11/12$.

1, it provides no information about the outcome of the white die. Example 2.5's argument (page 53) is pretty basic: $1/6^{th}$ of the time the red die is 1, and $1/6^{th}$ of *those* times the white die will also be 1. This is illustrated in Figure 2.12. Because the rolls are independent, the probabilities of the corresponding outcomes can be multiplied to get the final answer: $(1/6) * (1/6) = 1/36$. This can be generalized to many independent processes.

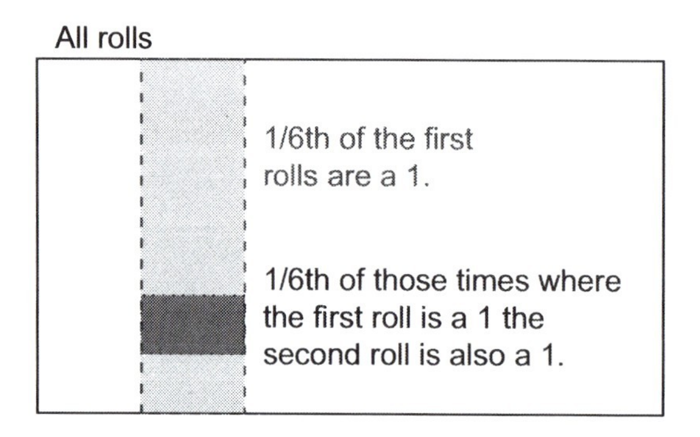

Figure 2.12: $1/6^{th}$ of the time, the first roll is a 1. Then $1/6^{th}$ of *those* times, the second roll will also be a 1.

● **Example 2.6** What if there was also a blue die independent of the other two? What is the probability of rolling the three dice and getting all 1s?

The same logic applies from Example 2.5. If $1/36^{th}$ of the time the white and red dice are both 1, then $1/6^{th}$ of **those** times the blue die will also be 1, so multiply:

$$\begin{aligned} &P(white = 1 \text{ and } red = 1 \text{ and } blue = 1) \\ &= P(white = 1) * P(red = 1) * P(blue = 1) \\ &= (1/6) * (1/6) * (1/6) = (1/36) * (1/6) = 1/216 \end{aligned}$$

Examples 2.5 and 2.6 illustrate what is called the Product Rule for independent processes.

Product Rule for independent processes

If A and B represent events from two different and independent processes, then the probability both A and B occur can be computed as the product of their separate probabilities:

$$P(A \text{ and } B) = P(A) * P(B)$$

Similarly, if there are k events A_1, ..., A_k from k independent processes, then the probability they all occur is

$$P(A_1) * P(A_2) * \cdots * P(A_k) \quad (2.7)$$

⊙ **Exercise 2.17** About 9% of people are `leftHanded`. Suppose 2 people are selected at random from the U.S. population. Because the sample of 2 is so small relative to the population, it is reasonable to assume these two people are independent. (a) What is the probability that both are `leftHanded`? (b) What is the probability that both are `rightHanded`? Answer to (a) and hint to (b) in the footnote[13].

⊙ **Exercise 2.18** Suppose 5 people are selected at random.

(a) What is the probability that all are `rightHanded`? Answer in the footnote[14].

(b) What is the probability that all are `leftHanded`?

(c) What is the probability that not all of the people are `rightHanded`? Hint in the footnote[15].

Suppose the variables `handedness` and `gender` are independent, i.e. knowing someone's `gender` provides no useful information about their `handedness` and vice-versa. We can compute whether a randomly selected person is `rightHanded` and `female` using the Product Rule:

$$\begin{aligned} P(\texttt{rightHanded} \text{ and } \texttt{female}) &= P(\texttt{rightHanded}) * P(\texttt{female}) \\ &= 0.91 * 0.50 = 0.455 \end{aligned}$$

The actual proportion of the U.S. population that is `female` is about 50%[16].

⊙ **Exercise 2.19** Three people are selected at random. (a) What is the probability the first person is `male` and `rightHanded`? Answer in the footnote[17]. (b) What is the probability the first two people are `male` and `rightHanded`?. (c) What is the probability the third person is `female` and `leftHanded`? (d) What is the probability the first two people are `male` and `rightHanded` and the third person is `female` and `leftHanded`? Short answers to (b)-(d) in the footnote[18].

---

[13](a) The probability the first person is `leftHanded` is 0.09, which is the same for the second person. We apply the Product Rule for independent processes to determine the probability that both will be `leftHanded`: $0.09 * 0.09 = 0.0081$.

(b) It is reasonable to assume the proportion of people who are ambidextrous (both right and left handed) is nearly 0, which results in $P(\texttt{rightHanded}) = 1 - 0.09 = 0.91$.

[14]Since each are independent, we apply the Product Rule for independent processes:

$$\begin{aligned} &P(\text{all five are } \texttt{rightHanded}) \\ &= P(\text{first} = \texttt{rH}, \text{second} = \texttt{rH}, ..., \text{fifth} = \texttt{rH}) \\ &= P(\text{first} = \texttt{rH}) * P(\text{second} = \texttt{rH}) * \cdots * P(\text{fifth} = \texttt{rH}) \\ &= 0.91 * 0.91 * 0.91 * 0.91 * 0.91 = 0.624 \end{aligned}$$

[15]Use the complement, $P(\text{all five are } \texttt{rightHanded})$, to answer this question.

[16]The proportion of men to women varies widely based on country.

[17]This is the same as $P(\text{a randomly selected person is } \texttt{male} \text{ and } \texttt{rightHanded}) = 0.455$.

[18](b) 0.207. (c) 0.045. (d) 0.0093.

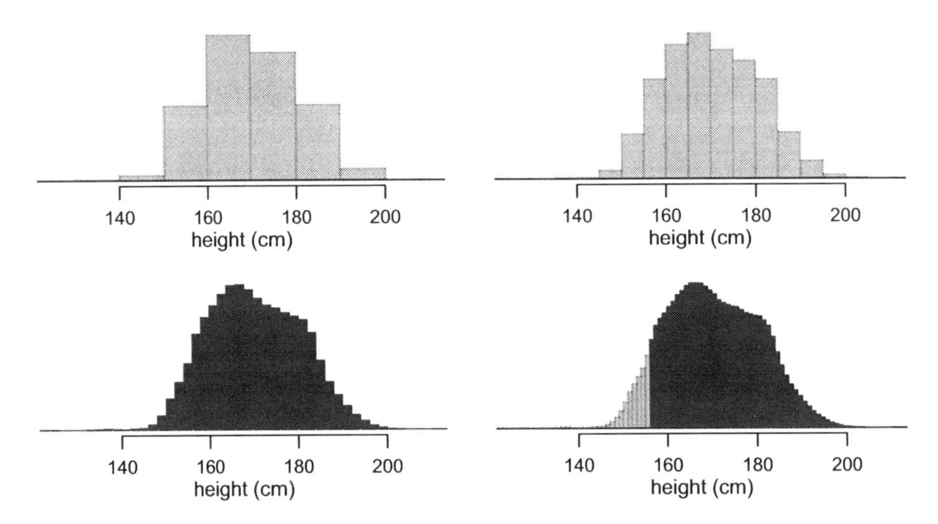

Figure 2.13: Four histograms of US adults heights with varying bin widths.

2.2 Continuous distributions

Example 2.7 Figure 2.13 shows a few different histograms of the variable `height` for 3 million US adults from the mid-90's[19]. How does changing the number of bins allow you to make different interpretations of the data?

While the histogram in the top-left panel has a few bins for a first rough approximation, more bins provide greater detail. This sample is extremely large, which is why much smaller bins still work well. (Usually we do not use so many bins with smaller sample sizes since small counts per bin mean the bin heights are very volatile.)

Example 2.8 What proportion of the sample is between 180 cm and 185 cm tall (about 5'11" to 6'1")?

The probability a randomly selected person is being between 180 cm and 185 cm can be estimated by finding the proportion of people in the sample who are between these two heights. In this case, this means adding up the heights of the bins in this range and dividing by the sample size. For instance, this can be done with the bins in the top right plot. The two bins in this region have counts of 195,307 and 156,239 people, resulting in the following estimate of the probability:

$$\frac{195307 + 156239}{3{,}000{,}000} = 0.1172$$

This fraction is the same as the proportion of the histogram's area that falls in the range 180 to 185 cm, shaded in Figure 2.14.

[19]This sample can be considered a simple random sample from the US population.

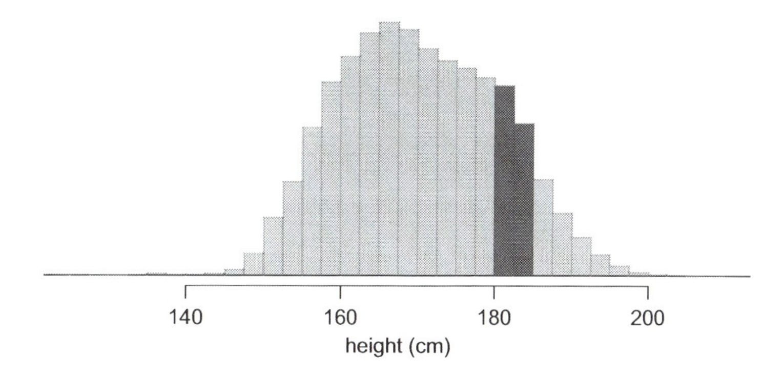

Figure 2.14: A histogram with bin sizes of 2.5 cm. The shaded region represents individuals with heights between 180 and 185 cm.

2.2.1 From histograms to continuous distributions

Examine the transition from a boxy histogram in the top-left of Figure 2.13 to the smooth histogram in the lower-right. In this last plot, the bins are so slim that the tops of the bins nearly create a smooth curve. This suggests the population height as a *continuous* numerical variable might best be explained by a curve that represents the top of extremely slim bins.

This smooth curve represents a **probability density function** (also called a **density** or **distribution**), and such a curve is shown in Figure 2.15 overlaid on a histogram of the sample. A density has a special property: the total area under the density's curve is 1.

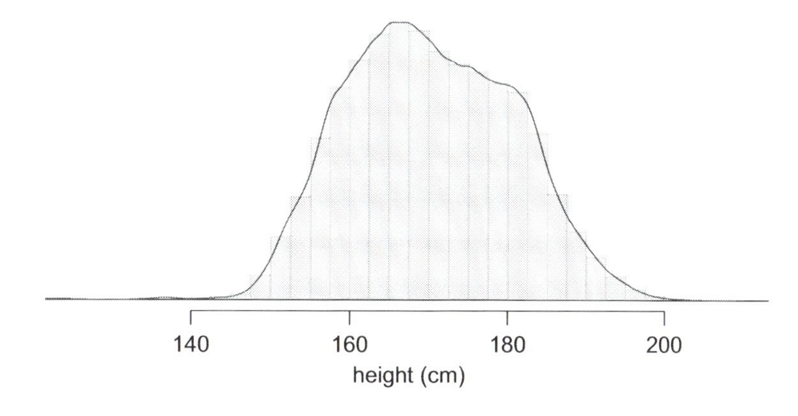

Figure 2.15: The continuous probability distribution of heights for US adults.

2.2.2 Probabilities from continuous distributions

We computed the proportion of individuals with heights 180 to 185 cm in Example 2.8 as a proportion:

$$\frac{\text{number of people between 180 and 185}}{\text{total sample size}}$$

We found the number of people with heights between 180 and 185 cm by determining the shaded boxes in this range, which represented the fraction of the box area in this region.

Similarly, we use the area in the shaded region under the curve to find a probability (with the help of a computer):

$$P(\texttt{height} \text{ between 180 and 185}) = \text{area between 180 and 185} = 0.1157$$

The probability a randomly selected person is between 180 and 185 cm is 0.1157. This is very close to the estimate from Example 2.8: 0.1172.

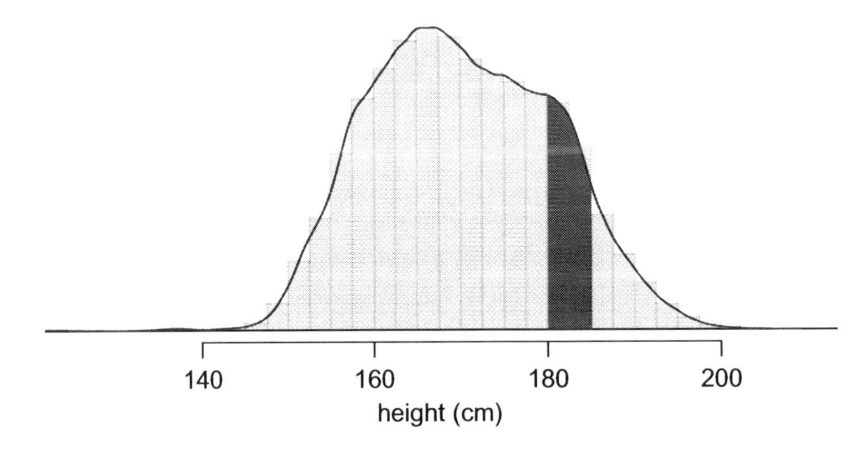

Figure 2.16: The total area under the curve representing all individuals is 1. The area between 180 and 185 cm is the fraction of the US adult population between 180 and 185 cm. Compare this plot with Figure 2.14.

⊙ **Exercise 2.20** Three US adults are randomly selected. The probability a single adult is between 180 and 185 cm is 0.1157. Short answers in the footnote[20].

(a) What is the probability that all three are between 180 and 185 cm tall?

(b) What is the probability that none are between 180 and 185 cm?

⊙ **Exercise 2.21** What is the probability a randomly selected person is **exactly** 180 cm? Assume you can measure perfectly. Answer in the footnote[21].

[20](a) $0.1157 * 0.1157 * 0.1157 = 0.0015$. (b) $(1 - 0.1157)^3 = 0.692$

[21]This probability is zero. While the person might be close to 180 cm, the probability that a randomly selected person is *exactly* 180 cm is zero. This also makes sense with the definition of probability as area; there is no area between 180 cm to 180 cm.

		parents		
		used	not	Total
student	uses	125	94	219
	not	85	141	226
	Total	210	235	445

Table 2.17: Contingency table summarizing the drugUse data set.

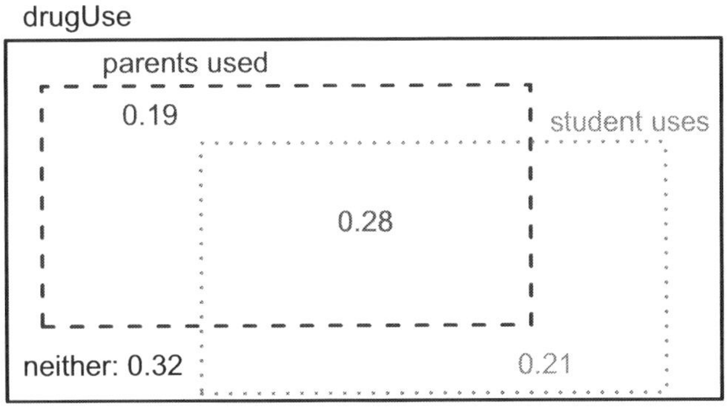

Figure 2.18: A Venn diagram using boxes for the drugUse data set.

⊙ **Exercise 2.22** Suppose a person's height is rounded to the nearest centimeter. Is there a chance that a random person's **measured** height will be 180 cm? Answer in the footnote[22].

2.3 Conditional probability

Are students more likely to use marijuana when their parents used drugs? The drugUse[23] data set contains a sample of 445 cases with two variables, student and parents, and is summarized in Table 2.17. The student variable is either uses or not, where a student uses if she has recently used marijuana. The parents variable takes the value used if at least one of the parents used drugs, including alcohol.

● **Example 2.9** If at least one parent used drugs, what is the chance their child (student) uses?

Of the 210 cases where parents was used, 125 represented cases where student = uses:

$$P(\texttt{student} = \texttt{uses} \text{ given } \texttt{parents} = \texttt{used}) = \frac{125}{210} = 0.60$$

[22]This has positive probability. Anyone between 179.5 cm and 180.5 cm will have a *measured* height of 180 cm. This is probably a more realistic scenario to encounter in practice versus Exercise 2.21.

[23]This data set is called parstum and is available in the faraway package for R.

	parents: used	parents: not	Total
student: uses	0.28	0.21	0.49
student: not	0.19	0.32	0.51
Total	0.47	0.53	1.000

Table 2.19: Probability table summarizing parental and student drug use.

For the cases in this data set, the probability is 0.60.

● **Example 2.10** A student is randomly selected from the study and she does not use drugs. What is the probability at least one of her parents used?

If the student does not use drugs, then she is one of the 226 students in the second row. Of these 226 students, 85 had parents who used drugs:

$$P(\texttt{parents} = \texttt{used} \text{ given } \texttt{student} = \texttt{not}) = \frac{85}{226} = 0.376$$

2.3.1 Marginal and joint probabilities

Table 2.17 includes row and column totals for the counts. These totals provide summary information about each variable separately.

We use these totals to compute **marginal probabilities** for the sample, which are the probabilities based on a single variable without conditioning on any other variables. For instance, a probability based solely on the `student` variable is a marginal probability:

$$P(\texttt{student} = \texttt{uses}) = \frac{219}{445} = 0.492$$

A probability of outcomes for two or more variables or processes is called a **joint probability**:

$$P(\texttt{student} = \texttt{uses} \ \& \ \texttt{parents} = \texttt{not}) = \frac{94}{445} = 0.21$$

It is common to substitute a comma for ampersand (&) in a joint probability, although either is acceptable.

Marginal and joint probabilities
If a probability is based on a single variable, it is a *marginal probability.* The probability of outcomes for two or more variables or processes is called a *joint probability.*

We use **table proportions** to summarize joint probabilities for the `drugUse` sample. These proportions are computed by dividing each count in Table 2.17 by 445 to obtain the proportions in Table 2.19. The joint probability distribution of the `parents` and `student` variables is shown in Table 2.20.

Joint outcome	Probability
parents = used, student = uses	0.28
parents = used, student = not	0.19
parents = not, student = uses	0.21
parents = not, student = not	0.32
Total	1.00

Table 2.20: A joint probability distribution for the drugUse data set.

⊙ **Exercise 2.23** Verify Table 2.20 represents a probability distribution: events are disjoint, all probabilities are non-negative, and the probabilities sum to 1.

⊙ **Exercise 2.24** Which table do you find more useful, Table 2.19 or Table 2.20? Why?

We can compute marginal probabilities based on joint probabilities in simple cases. The probability a random student from the study uses drugs is found by summing the outcomes from Table 2.20 where student = uses:

$$\begin{aligned} &P(\texttt{student} = \texttt{uses}) \\ &\quad = P(\texttt{parents} = \texttt{used}, \texttt{student} = \texttt{uses}) + \\ &\qquad\qquad P(\texttt{parents} = \texttt{not}, \texttt{student} = \texttt{uses}) \\ &\quad = 0.28 + 0.21 = 0.49 \end{aligned}$$

⊙ **Exercise 2.25** Let S and P represent student and parents, respectively. Then verify the equality holds in the following two equations:

$$\begin{aligned} P(\texttt{S} = \texttt{not}) &= P(\texttt{P} = \texttt{used} \;\&\; \texttt{S} = \texttt{not}) + P(\texttt{P} = \texttt{not} \;\&\; \texttt{S} = \texttt{not}) \\ P(\texttt{P} = \texttt{not}) &= P(\texttt{P} = \texttt{not} \;\&\; \texttt{S} = \texttt{uses}) + P(\texttt{P} = \texttt{not} \;\&\; \texttt{S} = \texttt{not}) \end{aligned}$$

⊙ **Exercise 2.26** Using Table 2.19, identify the probability at least one parent used drugs in a random case from the study?

2.3.2 Defining conditional probability

There is some connection between drug use of parents and of the student: drug use of one is associated with drug use of the other[24]. In this section, we discuss how to exploit associations between two variables to improve probability estimation.

The probability that a random student from the study uses drugs is 0.49. Could we update this probability if we knew that this student's parents used drugs? Absolutely. To do so, we limit our view to only those 210 cases where parents used drugs:

$$P(\texttt{student} = \texttt{uses given parents} = \texttt{used}) = \frac{125}{210} = 0.60$$

[24] This is an observational study and no causal conclusions may be reached.

We call this a **conditional probability** because we computed the probability under a condition: parents = used. There are two parts to a conditional probability, **the outcome of interest** and the **condition**. It is useful to think of the condition as information we know to be true, and this information usually can be described as a known outcome or event.

We separate the text inside our probability notation into the outcome of interest and the condition:

$$\begin{aligned} &P(\texttt{student} = \texttt{uses} \text{ given } \texttt{parents} = \texttt{used}) \\ &= P(\texttt{student} = \texttt{uses} \mid \texttt{parents} = \texttt{used}) = \frac{125}{210} = 0.60 \end{aligned} \tag{2.8}$$

The vertical bar "|" is read as *given*.

In Equation (2.8), we computed the probability a student uses based on the condition that at least one parent used as a fraction:

$$\begin{aligned} &P(\texttt{student} = \texttt{uses} \mid \texttt{parents} = \texttt{used}) \\ &= \frac{\#\text{ times } \texttt{student} = \texttt{uses} \;\&\; \texttt{parents} = \texttt{used}}{\#\text{ times } \texttt{parents} = \texttt{used}} \\ &= \frac{125}{210} = 0.60 \end{aligned} \tag{2.9}$$

We considered only those cases that met the condition, parents = used, and then we computed the ratio of those cases that satisfied our outcome of interest, that the student uses.

Counts are not always available for data, and instead only marginal and joint probabilities may be provided. Disease rates are commonly listed in percentages rather than in a raw data (count) format. We would like to be able to compute conditional probabilities even when no counts are available. We use Equation (2.9) as an example demonstrating this technique.

We considered only those cases that satisfied the condition, parents = used. Of these cases, the conditional probability was the fraction who represented the outcome of interest, student = uses. Suppose we were provided only the information in Table 2.19 on page 70, i.e. only probability data. Then if we took a sample of 1000 people, we would anticipate about 47% or $0.47*1000 = 470$ would meet our information criterion. Similarly, we would expect about 28% or $0.28*1000 = 280$ to meet both the information criterion and represent our outcome of interest. Thus, the conditional probability could be computed:

$$\begin{aligned} P(\texttt{S} = \texttt{uses} \mid \texttt{P} = \texttt{used}) &= \frac{\#\ (\texttt{S} = \texttt{uses} \;\&\; \texttt{P} = \texttt{used})}{\#\ (\texttt{P} = \texttt{used})} \\ &= \frac{280}{470} = \frac{0.28}{0.47} = 0.60 \end{aligned} \tag{2.10}$$

where S and P represent the student and parents variables. In Equation (2.10), we examine exactly the fraction of two probabilities, 0.28 and 0.47, which we can write as $P(\texttt{S} = \texttt{uses} \ \& \ \texttt{P} = \texttt{used})$ and $P(\texttt{P} = \texttt{used})$. The fraction of these probabilities represents our general formula for conditional probability.

Conditional Probability
The conditional probability of the outcome of interest A given condition B is computed as the following:

$$P(A|B) = \frac{P(A \ \& \ B)}{P(B)} \tag{2.11}$$

⊙ **Exercise 2.27** (a) Write out the following statement in conditional probability notation: "*The probability a random case has* `parents` = `not` *if it is known that* `student` = `not`". Notice that the condition is now based on the student, not the parent. (b) Determine the probability from part (a). Table 2.19 on page 70 may be helpful. Answers in the footnote[25].

⊙ **Exercise 2.28** (a) Determine the probability that one of the parents had used drugs if it is known the student does not use drugs, i.e.

$$P(\texttt{parents} = \texttt{used} \ |\texttt{student} = \texttt{not})$$

(b) Using part (a) and Exercise 2.27(b), compute

$$P(\texttt{parents} = \texttt{used}|\texttt{student} = \texttt{not}) + P(\texttt{parents} = \texttt{not}|\texttt{student} = \texttt{not})$$

(c) Provide an intuitive argument to explain why the sum in (b) is 1. Answer for part (c) in the footnote[26].

⊙ **Exercise 2.29** It has been found in this data that drug use between parents and children tends to be associated. Does this mean the drug use of parents causes the drug use of the students? Answer in the footnote[27].

2.3.3 Smallpox in Boston, 1721

The `smallpox` data set provides a sample of 10,700 individuals from the year 1721 who were potentially exposed to smallpox in Boston. Doctors at the time believed that inoculation, which involves exposing a person to the disease in a controlled form, could reduce the likelihood of death.

Each case represents one person with two variables: `inoculated` and `result`. The variable `inoculated` takes two levels: `yes` or `no`, indicating whether the person

[25](a) $P(\texttt{parent} = \texttt{not}|\texttt{student} = \texttt{not})$. (b) Equation (2.11) for conditional probability suggests we should first find $P(\texttt{parents} = \texttt{not} \ \& \ \texttt{student} = \texttt{not}) = 0.32$ and $P(\texttt{student} = \texttt{not}) = 0.51$. Then the ratio represents the conditional probability: $0.32/0.51 = 0.63$.

[26]Under the condition the student does not use drugs, the parents must either use drugs or not. The complement still appears to work *when conditioning on the same information.*

[27]No. This was an observational study. Two potential lurking variables include `income` and `region`. Can you think of others?

		inoculated		
		yes	no	Total
result	lived	281	9571	9852
	died	6	842	848
	Total	287	10,413	10,700

Table 2.21: Contingency table for the `smallpox` data set.

		inoculated		
		yes	no	Total
result	lived	0.0263	0.8945	0.9207
	died	0.0006	0.0787	0.0793
	Total	0.0268	0.9732	1.0000

Table 2.22: Table proportions for the `smallpox` data, computed by dividing each count by the table total, 10,700. Some totals are slightly off due to rounding errors.

was inoculated or not. The variable `result` has outcomes `lived` or `died`. We summarize the data in Table 2.21. Table 2.22 displays the results using table proportions.

⊙ **Exercise 2.30** (a) Write out, in formal notation, the probability a randomly selected person who was not inoculated died from smallpox. (b) Find the probability in part (a). Brief answers in the footnote[28].

⊙ **Exercise 2.31** (a) Determine the probability an inoculated person died from smallpox. (b) How does this result compare with the result of Exercise 2.30?

⊙ **Exercise 2.32** The people of Boston self-selected whether or not to be inoculated. (a) Is this study observational or experimental? (b) Can we infer a causal connection between inoculation and whether someone died from smallpox? Answers in the footnote[29].

⊙ **Exercise 2.33** What are some potential lurking variables that might influence whether someone `lived` or `died` that might also affect whether she was inoculated or not? One is provided in the footnote[30].

2.3.4 General multiplication rule

Section 2.1.5 introduced a multiplication rule for independent events. Here we provide a General Multiplication Rule for events that might not be independent.

[28] $P(\texttt{result} = \texttt{died} \mid \texttt{inoculated} = \texttt{no}) = \frac{0.0787}{0.9732} = 0.0809$.

[29] (a) Observational. (b) No! We cannot infer causation from this observational study.

[30] Accessibility to the latest and best medical care.

General Multiplication Rule
If A and B represent two outcomes or events, then

$$P(A \text{ and } B) = P(A|B)P(B)$$

It might be useful to think of A as the outcome of interest and B as the condition.

This General Multiplication Rule is simply a rearrangement of the definition for conditional probability in Equation (2.11) on page 73.

● **Example 2.11** Consider the `smallpox` data set. Suppose we only knew that 91.91% of those not inoculated survived, and we also knew that 97.32% of all individuals were not inoculated. How could we compute the probability someone was not inoculated and also survived?

We will compute our answer and then verify it using Table 2.22. We want to determine

$$P(\texttt{result} = \texttt{lived} \ \& \ \texttt{inoculated} = \texttt{no})$$

and we are given that

$$P(\texttt{result} = \texttt{lived} \mid \texttt{inoculated} = \texttt{no}) = 0.9191$$
$$P(\texttt{inoculated} = \texttt{no}) = 0.9732$$

Of the 97.32% of people who were not inoculated, 91.91% survived:

$$P(\texttt{result} = \texttt{lived} \ \& \ \texttt{inoculated} = \texttt{no}) = 0.9191 * 0.9732 = 0.8945$$

This is equivalent to the General Multiplication Rule, and we can confirm this probability in Table 2.22 on the facing page at the intersection of `no` and `lived`.

⊙ **Exercise 2.34** Use $P(\texttt{inoculated} = \texttt{yes}) = 0.0268$ and $P(\texttt{result} = \texttt{lived} \mid \texttt{inoculated} = \texttt{yes}) = 0.9791$ to determine the probability a person was both inoculated and lived. Verify your answer in Table 2.22. Your answer will be off by 0.0001 due to rounding error.

Sum of conditional probabilities
Let A_1, ..., A_k represent all the disjoint outcomes for a variable or process. Then if B is a condition for another variable or process, we have:

$$P(A_1|B) + \cdots + P(A_k|B) = 1$$

The rule for complements also holds when an event and its complement are conditioned on the same information:

$$P(A|B) = 1 - P(A^c|B)$$

⊙ **Exercise 2.35** If 97.91% of the people who were inoculated lived, what proportion of inoculated people must have died?

2.3.5 Independence considerations

If two processes are independent, then knowing the outcome of one should have no influence on the other. We can show this is mathematically true using conditional probabilities.

⊙ **Exercise 2.36** Let X and Y represent the outcomes of rolling two dice. (a) What is the probability the first die, X, is `1`? (b) What is the probability both X and Y are `1`? (c) Use the formula for conditional probability to compute $P(Y = \texttt{1} \,|X = \texttt{1})$. (d) If X was known to be `1`, did it alter the probability Y was `1` in part (c)?

We can show in Exercise 2.36(c) that the conditioning information has no influence by using the Multiplication Rule for independence processes:

$$\begin{aligned} P(X = \texttt{1}|Y = \texttt{1}) &= \frac{P(X = \texttt{1} \ \&\ Y = \texttt{1})}{P(Y = \texttt{1})} \\ &= \frac{P(X = \texttt{1}) * P(Y = \texttt{1})}{P(Y = \texttt{1})} \\ &= P(X = \texttt{1}) \end{aligned}$$

⊙ **Exercise 2.37** Ron is watching a roulette table in a casino and notices that the last five outcomes were `black`. He figures that the chances of getting `black` six times in a row is very small (1/64 to be exact) and puts his paycheck on red. What is wrong with his reasoning?

2.3.6 Tree diagrams

Tree diagrams are a tool to organize outcomes and probabilities around the structure of the data. They are most useful when two or more processes occur in a sequence and each process is conditioned on its predecessors.

The `smallpox` data fits this description. We see the population as split by `inoculation`: `yes` and `no`. Following this split, survival rates were observed for each group. We construct the tree diagram to follow the data structure. It splits the group by `inoculation` and then split those subgroups by `result`, shown in Figure 2.23. The first branch for `inoculation` is said to be the **primary** branch while the other branches are **secondary**.

We annotate the tree diagram with marginal and conditional probabilities[31]. We first split the data by `inoculation` into the `yes` and `no` groups with respective marginal probabilities 0.0268 and 0.9732. The second split is conditioned on the

[31]Marginal and joint probabilities were discussed in Section 2.3.1.

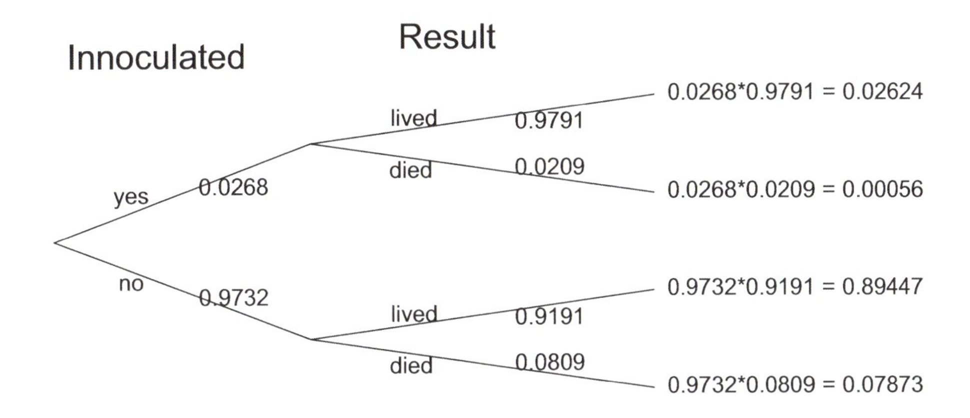

Figure 2.23: A tree diagram of the `smallpox` data set.

first, so we assign conditional probabilities to the branches. For example, the top branch in Figure 2.23 is the probability that `result` = `lived` conditioned on the information that `inoculated` = `yes`. We may (and usually do) construct joint probabilities at the end of each branch in our tree by multiplying the numbers we come across as we move from left to right. These joint probabilities are computed directly from the General Multiplication Rule:

$$\begin{aligned}&P(\texttt{inoculated} = \texttt{yes}\ \&\ \texttt{result} = \texttt{lived})\\ &= P(\texttt{inoculated} = \texttt{yes}) * P(\texttt{result} = \texttt{lived}|\texttt{inoculated} = \texttt{yes})\\ &= 0.0268 * 0.9791 = 0.02624\end{aligned}$$

● **Example 2.12** Consider the `midterm` and `final` for a statistics class. Suppose 13% of students earn an `A` on the `midterm`. Of those students who earned an `A` on the `midterm`, 47% got an `A` on the `final`, and 11% of the students who got lower than an `A` on the `midterm` got an `A` on the `final`. You randomly pick up a final exam and notice the student got an `A`. What is the probability this student got an `A` on the `midterm`?

It is not obvious how to solve this problem. However, we can start by organizing our information into a tree diagram. First split the students based on `midterm` and then split those primary branches into secondary branches based on `final`. We associate the marginal probability with the primary branching and the conditional probabilities with the secondary branches. The result is shown in Figure 2.24.

The end-goal is to find $P(\texttt{midterm} = \texttt{A}|\texttt{final} = \texttt{A})$. We can start by finding the two probabilities associated with this conditional probability using the tree diagram:

$$\begin{aligned}&P(\texttt{final} = \texttt{A}\ \&\ \texttt{midterm} = \texttt{A}) = 0.0611\\ &P(\texttt{final} = \texttt{A})\\ &\quad = P(\texttt{final} = \texttt{A}\ \&\ \texttt{midterm} = \texttt{other}) + P(\texttt{final} = \texttt{A}\ \&\ \texttt{midterm} = \texttt{A})\\ &\quad = 0.0611 + 0.0957 = 0.1568\end{aligned}$$

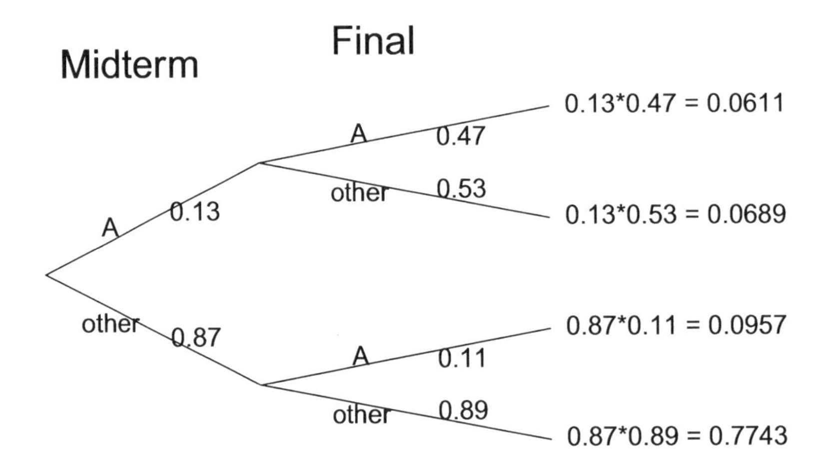

Figure 2.24: A tree diagram describing the `midterm` and `final` variables.

Then we take the ratio of these two probabilities:

$$P(\texttt{midterm} = \texttt{A}|\texttt{final} = \texttt{A}) = \frac{P(\texttt{midterm} = \texttt{A}\ \&\ \texttt{final} = \texttt{A})}{P(\texttt{final} = \texttt{A})} = \frac{0.0611}{0.1568} = 0.3897$$

The probability the student also got an A on the midterm is about 0.39.

⊙ **Exercise 2.38** After an introductory statistics course, 78% of students can successfully construct tree diagrams. Of those who can construct tree diagrams, 97% passed, while only 57% of those students who could not construct tree diagrams passed. (a) Organize this information into a tree diagram. (b) What is the probability a randomly selected student passed? (c) Compute the probability a student is able to construct a tree diagram if it is known that she passed. Hints plus a short answer to (c) in the footnote[32].

2.4 Sampling from a small population

● **Example 2.13** Professors sometimes select a student at random to answer a question. If the selection is truly random, and there are 15 people in your class, what is the chance that she will pick you for the next question?

If there are 15 people to ask and none are skipping class (for once), then the probability is 1/15, or about 0.067.

[32](a) The first branch should be based on the variable that directly splits the data into groups: whether students can construct tree diagrams. (b) Identify which two joint probabilities represent students who passed. (c) Use the definition of conditional probability. Your answer to part (b) may be useful in part (c). The solution to (c) is 0.86.

● **Example 2.14** If the professor asks 3 questions, what is the probability you will not be selected? Assume that she will not pick the same person twice in a given lecture.

For the first question, she will pick someone else with probability 14/15. When she asks the second question, she only has 14 people who have not yet been asked. Thus, if you were not picked on the first question, the probability you are again not picked is 13/14. Similarly, the probability you are again not picked on the third question is 12/13, and the probability of not being picked for any of the three questions is

$$\begin{aligned} &P(\text{not picked in 3 questions}) \\ &\quad = P(\texttt{Q1} = \texttt{notPicked}, \texttt{Q2} = \texttt{notPicked}, \texttt{Q3} = \texttt{notPicked}.) \\ &\quad = \frac{14}{15} \times \frac{13}{14} \times \frac{12}{13} = \frac{12}{15} = 0.80 \end{aligned}$$

⊙ **Exercise 2.39** What rule permitted us to multiply the probabilities in Example 2.14? Answer in the footnote[33]

● **Example 2.15** Suppose the professor randomly picks without regard to who she already picked, i.e. students can be picked more than once. What is the probability you will not be picked for any of the three questions?

Each pick is independent, and the probability of `notPicked` on any one question is 14/15. Thus, we can use the Product Rule for independent processes.

$$\begin{aligned} &P(\text{not picked in 3 questions}) \\ &\quad = P(\texttt{Q1} = \texttt{notPicked}, \texttt{Q2} = \texttt{notPicked}, \texttt{Q3} = \texttt{notPicked}.) \\ &\quad = \frac{14}{15} \times \frac{14}{15} \times \frac{14}{15} = 0.813 \end{aligned}$$

You have a slightly higher chance of not being picked compared to when she picked a new person for each question. However, you now may be picked more than once.

⊙ **Exercise 2.40** Under the setup of Example 2.15, what is the probability of being picked to answer all three questions?

If we sample from a small population **without replacement**, we no longer have independence between our observations. In Example 2.14, the probability of `notPicked` for the second question was conditioned on the event that you were `notPicked` for the first question. In Example 2.15, the professor sampled her students **with replacement**: she repeatedly sampled the entire class without regard to who she already picked.

---

[33]The three probabilities we computed were actually one marginal probability, $P(\texttt{Q1}=\texttt{notPicked})$, and two conditional probabilities:

$$P(\texttt{Q2} = \texttt{notPicked.}|\texttt{Q1} = \texttt{notPicked.})$$
$$P(\texttt{Q3} = \texttt{notPicked.}|\texttt{Q1} = \texttt{notPicked.}\ \texttt{Q2} = \texttt{notPicked.})$$

Using the General Multiplication Rule, the product of these three probabilities is the probability of not being picked in 3 questions.

⊙ **Exercise 2.41** Your department is holding a raffle. They sell 30 tickets and offer seven prizes. (a) They place the tickets in a hat and draw one for each prize. The tickets are sampled without replacement, i.e. the selected tickets are not placed back in the hat. What is the probability of winning a prize if you buy one ticket? (b) What if the tickets are sampled with replacement? Answers in the footnote[34].

⊙ **Exercise 2.42** Compare your answers in Exercise 2.41. How much influence does the sampling method have on your chances of winning a prize?

Had we repeated Exercise 2.41 with 300 tickets instead of 30, we would have found something interesting: the results would be nearly identical. The probability would be 0.0233 without replacement and 0.0231 with replacement. When the sample size is only a small fraction of the population (under 10%), observations are nearly independent even when sampling without replacement.

2.5 Expected value and uncertainty

● **Example 2.16** Two books are assigned for a statistics class: a textbook and its corresponding study guide. The student bookstore determined 20% of enrolled students do not buy either book, 55% buy the textbook, and 25% buy both books, and these percentages are relatively constant from one term to another. If there are 100 students enrolled, how many books should the bookstore expect to sell to this class?

Around 20 students will not buy either book (0 books total), about 55 will buy one book (55 total), and approximately 25 will buy two books (totaling 50 books for these 25 students). The bookstore should expect to sell about 105 books for this class.

⊙ **Exercise 2.43** Would you be surprised if the bookstore sold slightly more or less than 105 books?

● **Example 2.17** The textbook costs \$137 and the study guide \$33. How much revenue should the bookstore expect from this class of 100 students?

About 55 students will just buy a textbook, providing revenue of

$$\$137 * 55 = \$7,535$$

[34] (a) First determine the probability of not winning. The tickets are sampled without replacement, which means the probability you do not win on the first draw is 29/30, 28/29 for the second, ..., and 23/24 for the seventh. The probability you win no prize is the product of these separate probabilities: 23/30. That is, the probability of winning a prize is $7/30 = 0.233$. (b) When the tickets are sampled with replacement, they are seven independent draws. Again we first find the probability of not winning a prize: $(29/30)^7 = 0.789$. Thus, the probability of winning (at least) one prize when drawing with replacement is 0.211.

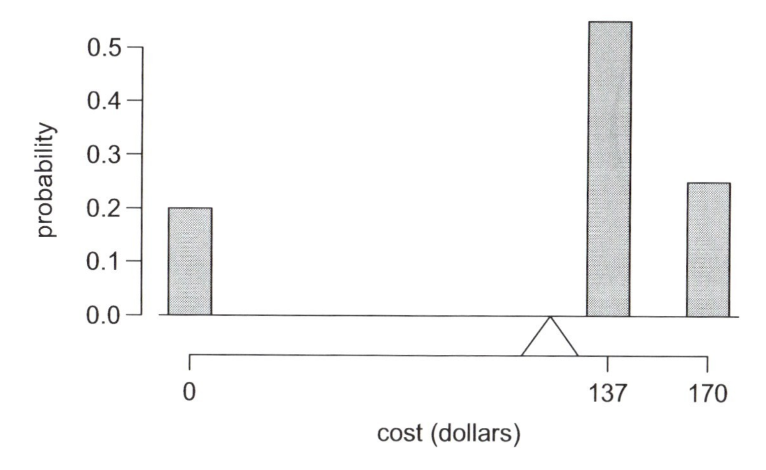

Figure 2.25: Probability distribution for the bookstore's revenue from a single student. The distribution balances on a pyramid representing the average revenue per student.

The roughly 25 students who buy both the textbook and a study guide would pay a total of

$$(\$137 + \$33) * 25 = \$170 * 25 = \$4,250$$

Thus, the bookstore should expect to generate about \$11,785 from these 100 students for this one class. However, there might be some *sampling variability* so the actual amount may differ by a little bit.

● **Example 2.18** What is the average revenue per student for this course?

The expected total revenue is \$11,785, and there are 100 students. Therefore the expected revenue per student is $\$11,785/100 = \$117.85$.

2.5.1 Expectation

We call a variable or process with a numerical outcome a **random variable**, and usually represent this random variable with a capital letter such as X, Y, or Z. The amount of money a single student will spend on her statistics books is a random variable, and we represent its outcome by X.

Random variables
A random process or variable with a numerical outcome.

The possible outcomes of X are labeled with a corresponding lower case letter with subscripts. For example, we write $x_1 = \$0$, $x_2 = \$137$, and $x_3 = \$170$, which occur with probabilities 0.20, 0.55, and 0.25. The distribution of X is summarized in Figure 2.25 and Table 2.26.

i	1	2	3	Total
x_i	\$0	\$137	\$170	–
$P(X = x_i)$	0.20	0.55	0.25	1.00

Table 2.26: The probability distribution for the random variable X, representing the bookstore's revenue from a single student.

We computed the average outcome of X as \$117.85 in Example 2.18. We call this average the **expected value** of X, denoted by $E(X)$. The expected value of a random variable is computed by adding each outcome weighted by its probability:

$$\begin{aligned} E(X) &= \$0 * P(X = \$0) + \$137 * P(X = \$137) + \$170 * P(X = \$170) \\ &= \$0 * 0.20 + \$137 * 0.55 + \$170 * 0.25 = \$117.85 \end{aligned}$$

Expected value of a Discrete Random Variable
If X takes outcomes x_1, ..., x_k with probabilities $P(X = x_1)$, ..., $P(X = x_k)$, the expected value of X is the sum of each outcome multiplied by its corresponding probability:

$$\begin{aligned} E(X) &= x_1 * P(X = x_1) + \cdots + x_k * P(X = x_k) \\ &= \sum_{i=1}^{k} x_i P(X = x_i) \end{aligned} \qquad (2.12)$$

The notation μ may be used in place of $E(X)$.

The expected value for a random variable represents the average outcome. For example, $E(X) = 117.85$ represents the average amount the bookstore expects to make from a single student, and so we could write it as $\mu = 117.85$.

It is also possible to compute the expected value of a continuous random variable. However, it requires a little calculus and we save it for a later class[35].

In physics, the expectation holds the same meaning as a center of gravity. The distribution can be represented by a series of weights at each outcome, and the mean represents the balancing point. This is represented in Figures 2.25 and 2.27. The idea of a center of gravity also expands to a continuous probability distribution. Figure 2.28 shows a continuous probability distribution balanced atop a wedge placed at the mean.

2.5.2 Variability in random variables

Suppose you ran the university bookstore. What else would you like to know besides how much revenue you expect to generate? You might also want to know

[35] $\mu = \int x P(X = x) dx$

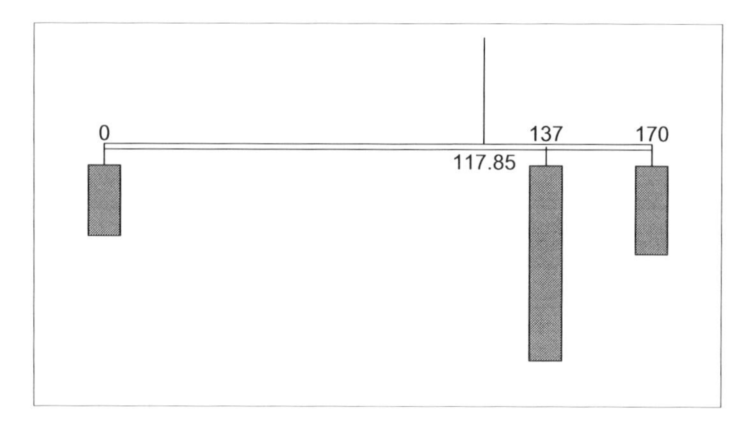

Figure 2.27: A weight system representing the probability distribution for X. The string holds the distribution at the mean to keep the system balanced.

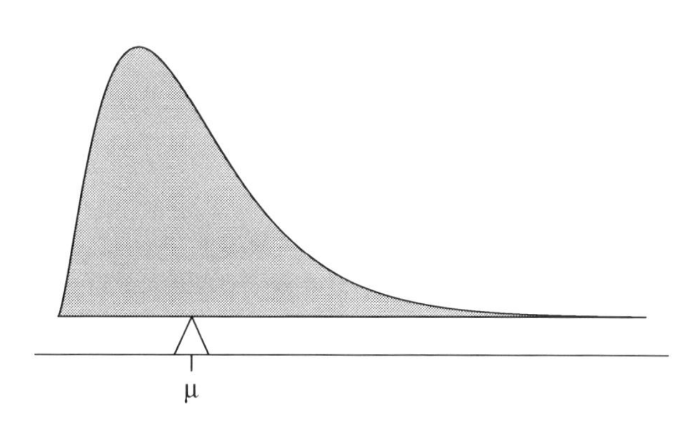

Figure 2.28: A continuous distribution can also be balanced at its mean.

the volatility (variability) in your revenue.

The variance and standard deviation can be used to describe the variability of a random variable. Section 1.3.5 introduced a method for finding the variance and standard deviation for a data set. We first computed deviations from the mean ($x_i - \mu$), squared those deviations, and took an average to get the variance. In the case of a random variable, we again compute squared deviations. However, we take their sum weighted by their corresponding probabilities, just like we did for expectation. This weighted sum of squared deviations equals the variance, and we calculate the standard deviation by taking the square root, just as we did in Section 1.3.5.

General variance formula
If X takes outcomes x_1, ..., x_k with probabilities $P(X = x_1)$, ..., $P(X = x_k)$ and $\mu = E(X)$, then the variance of X, denoted by $Var(X)$ or the symbol σ^2, is

$$\sigma^2 = (x_1 - \mu)^2 * P(X = x_1) + \cdots$$
$$\cdots + (x_k - \mu)^2 * P(X = x_k)$$
$$= \sum_{j=1}^{k} (x_j - \mu)^2 P(X = x_j) \quad (2.13)$$

The standard deviation of X, σ, is the square root of the variance.

● **Example 2.19** Compute the expected value, variance, and standard deviation of X, the revenue of a single statistics student for the bookstore.

It is useful to construct a table that holds computations for each outcome separately, then add them up.

i	1	2	3	Total
x_i	\$0	\$137	\$170	
$P(X = x_i)$	0.20	0.55	0.25	
$x_i * P(X = x_i)$	0	75.35	42.50	117.85

Thus, the expected value is $\mu = 117.35$, which we computed earlier. The variance can be constructed by extending this table:

i	1	2	3	Total
x_i	\$0	\$137	\$170	
$P(X = x_i)$	0.20	0.55	0.25	
$x_i * P(X = x_i)$	0	75.35	42.50	117.85
$x_i - \mu$	-117.85	19.15	52.15	
$(x_i - \mu)^2$	13888.62	366.72	2719.62	
$(x_i - \mu)^2 * P(X = x_i)$	2777.7	201.7	679.9	3659.3

The variance of X is $\sigma^2 = 3659.3$, which means the standard deviation is $\sigma = \sqrt{3659.3} = \$60.49$.

⊙ **Exercise 2.44** The bookstore also offers a chemistry textbook for \$159 and a book supplement for \$41. From past experience, they know about 25% of chemistry students just buy the textbook while 60% buy both the textbook and supplement. Answers for each part below are provided in the footnote[36].

(a) What proportion of students don't buy either book? Again assume no students buy the supplement without the textbook.

(b) Let Y represent the revenue from a single student. Write out the probability distribution of Y, i.e. a table for each outcome and its associated probability.

(c) Compute the expected revenue from a single chemistry student.

(d) Find the standard deviation to describe the variability associated with the revenue from a single student.

[36](a) 100% - 25% - 60% = 15% of students do not buy any books for the class. Part (b) is represented by lines Y and $P(Y)$ below. The expectation for part (c) is given as the total on the line $X * P(X)$. The result of part (d) is the square-root of the variance listed on in the total on the last line: $\sigma = \sqrt{Var(Y)} = \$69.28$.

Scenario	noBook	textbook	both	Total
Y	0.00	159.00	200.00	
$P(Y)$	0.15	0.25	0.60	
$Y * P(Y)$ 3	0.00	39.75	120.00	$E(Y) = \$159.75$
$Y - E(Y)$	-159.75	-0.75	40.25	
$(Y - E(Y))^2$	25520.06	0.56	1620.06	
$(Y - E(Y))^2 * P(Y)$	3828.0	0.1	972.0	$Var(Y) \approx 4800$

2.6 Problem set

2.6.1 Defining probability

2.1 Answer the following questions.

(a) If a fair coin is tossed many times and the last eight tosses are all heads, then the chance that the next toss will be heads is somewhat less than 50%. True or false and justify.

(b) A coin is tossed and you win a prize if there are more than 60% heads. Would you be surprised if the coin landed on heads more than 60% of the time in 10 heads? in 100 heads? Which is more surprising? Which is better in order to win the prize: 10 tosses or 100 tosses? Explain.

(c) A coin is tossed and you win a prize if there are more than 40% heads. Which is better: 10 tosses or 100 tosses? Explain.

(d) A coin is tossed and you win a prize if there are between 40% and 60% heads. Which is better: 10 tosses or 100 tosses? Explain.

2.2 Backgammon is a board game for two players in which the playing pieces are moved according to the roll of two dice. Players win by removing all of their pieces from the board therefore it is good to roll high numbers. You are playing backgammon with a friend and you roll two double sixes in a row while your friend rolled first double ones and then double threes. Double sixes means that in a single roll of two dice, each die comes up as a six. Double threes means each die comes up as a three, etc. Your friend claims that you are cheating because it is very unlikely to be able to roll two double sixes in a row. Using probabilities how could you argue that your rolls were just as likely as his?

2.3 If you flip a fair coin 10 times, what is the probability of

(a) getting all tails?

(b) getting all heads?

(c) getting at least one tail?

2.4 In a class where everyone's native language is English, 70% of students speak Spanish, 45% speak French as a second language and 20% speak both. Assume that there are no students who speak another second language.

(a) What percent of the students do not speak a second language?

(b) Are speaking French and Spanish mutually exclusive?

(c) Are the variables speaking French and speaking Spanish independent?

2.5 In a multiple choice exam there are 5 questions and 4 choices for each question (a, b, c, d). Nancy has not studied for the exam at all, and decided to randomly guess the answers. What is the probability that:

(a) The first question she gets right is the 5th question? $(\frac{3}{4})^4 (\frac{1}{4})$

(b) She gets all questions right? $(\frac{1}{4})^5$

(c) She gets at least one question right? ---- $1 -$ all wrong $= 1 - (\frac{3}{4})^5$

2.6 The US Census is conducted every 10 years and collects demographic information from the residents of United States. The table below shows the distribution of the level highest education obtained by US residents by gender. Answer the following questions based on this table.

		Gender	
		Male	Female
	Less than high school	0.19	0.19
	High school graduate	0.28	0.30
Highest	Some college	0.27	0.28
education	Bachelor's degree	0.16	0.15
attained	Master's degree	0.06	0.06
	Professional school degree	0.03	0.01
	Doctorate degree	0.01	0.01

(a) What is the probability that a randomly chosen man has at least a Bachelor's degree?

(b) What is the probability that a randomly chosen woman has at least a Bachelor's degree?

(c) What is the probability that a man and a woman getting married both have at least a Bachelor's degree?

(d) What assumption did you make to calculate the probability in part (d)? Do you think this assumption is reasonable?

2.7 Indicate what, if anything, is wrong with the following probability distributions for grade breakdown in a Statistics class.

	Grades				
	A	B	C	D	F
(a)	0.3	0.3	0.3	0.2	0.1
(b)	0	0	1	0	0
(c)	0.3	0.3	0.3	0.1	0
(d)	0.3	0.5	0.2	0.1	-0.1
(e)	0.2	0.4	0.2	0.1	0.1
(f)	0	-0.1	1.1	0	0

2.8 The table below shows the relationship between hair color and eye color for a group of 1,770 German men.

		Hair Color			
		Brown	Black	Red	Total
Eye	Brown	400	300	20	720
Color	Blue	800	200	50	1050
	Total	1200	500	70	1770

(a) If we draw one man at random, what is the probability that he has brown hair and blue eyes?

(b) If we draw one man at random, what is the probability that he has brown hair or blue eyes?

2.6.2 Continuous distributions

2.9 Below is a histogram of body weights (in *kg*) of 144 male and female cats.

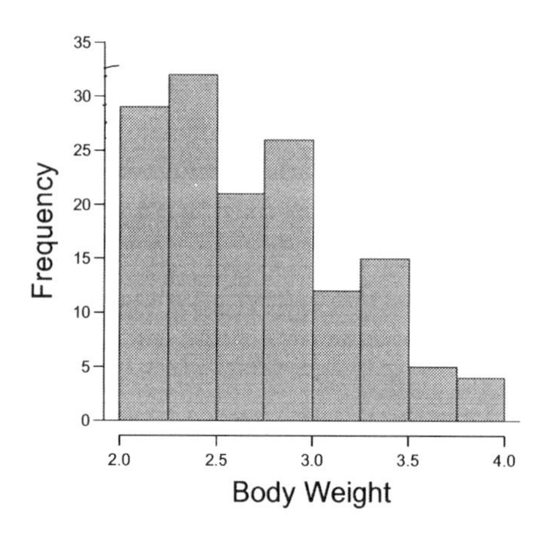

(a) What is the probability that a randomly chosen cat weighs less than 2.5 *kg*?

(b) What is the probability that a randomly chosen cat weighs between 2.5 and 2.75 *kg*?

(c) What is the probability that a randomly chosen cat weighs between 3 and 3.5 *kg*?

(d) What is the probability that a randomly chosen cat weighs more than 3.5 *kg*?

2.10 Exercise 9 introduces a data set of 144 cats' body weights. The below histogram shows the distribution of these cats' body weights by gender. There are 47 female and 97 male cats in the data set.

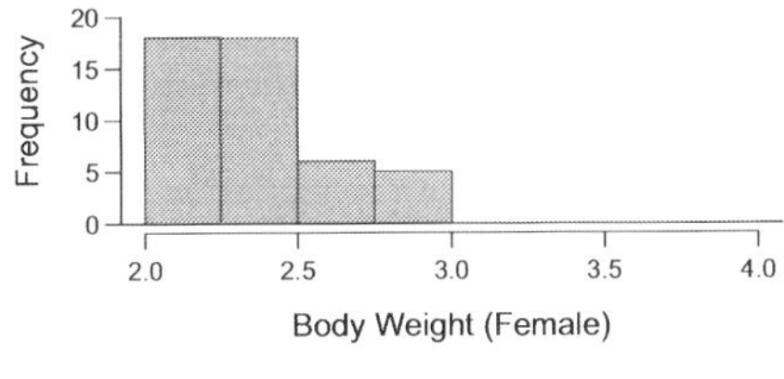

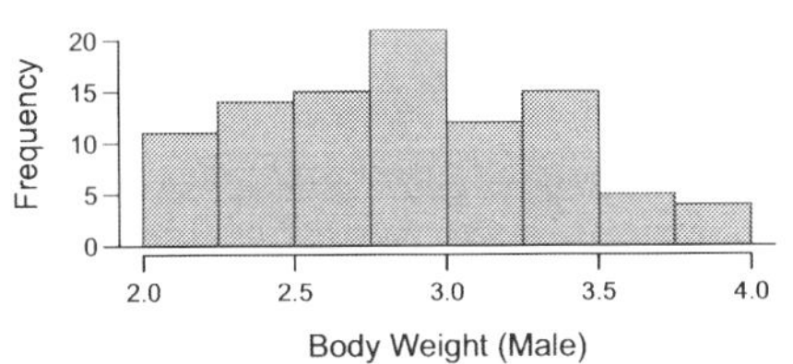

(a) What is the probability that a randomly chosen female cat weighs less than 2.5 kg?

(b) What is the probability that a randomly chosen male cat weighs less than 2.5 kg?

2.11 The US Census is conducted every 10 years and collects demographic information from the residents of United States. Below are a histogram representing the distribution of total personal income of a random sample of 391 people from the 2000 Census and a frequency table based on which the histogram was constructed. Also shown is the distribution of gender in the sample and the distribution of income for males and females.

(a) Describe the distribution of total personal income in this sample.

(b) What is the probability that a randomly chosen person makes less than \$50,000 per year?

(c) What is the probability that a randomly chosen person is female?

(d) Assuming that total personal income and gender are independent, what is the probability that a randomly chosen person makes less than \$50,000 per year and is female?

(e) Do you think the assumption made in part (d) is reasonable? Use the box plot to justify your answer.

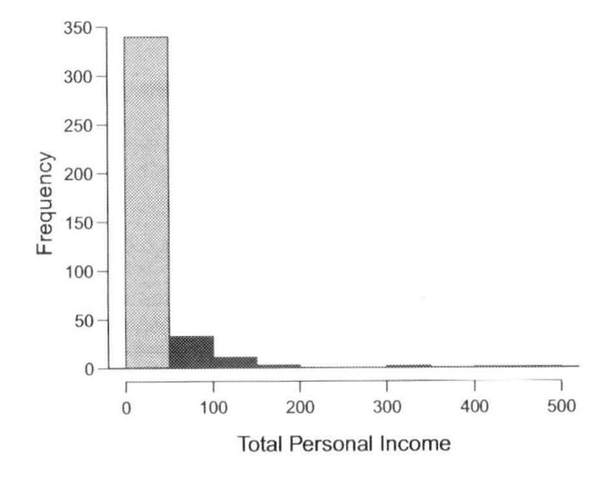

Income	*Frequency*
0 - 50,000	340
50,001 - 100,000	33
100,001 - 150,000	11
150,001 - 200,000	3
200,001 - 250,000	0
250,001 - 300,000	0
300,001 - 350,000	2
350,001 - 400,000	0
400,001 - 450,000	1
450,001 - 500,000	1

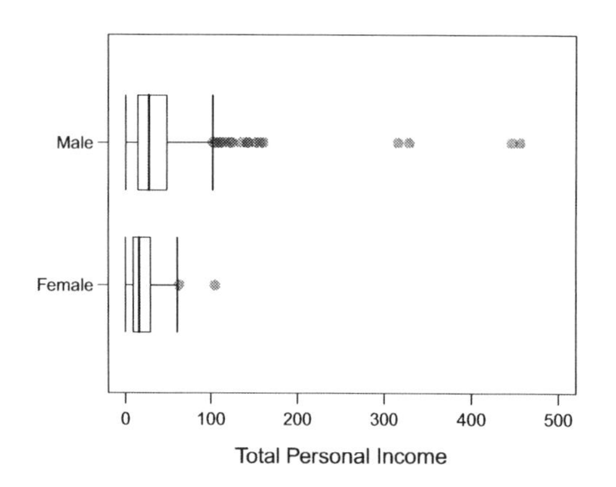

Gender	Frequency
Female	188
Male	203

2.6.3 Conditional probability

2.12 P(A) = 0.3, P(B) = 0.7

(a) Can you compute P(A and B) if you only know P(A) and P(B)?

(b) Assuming that events A and B arise from independent random processes,

  (i) what is P(A and B)?

  (ii) what is P(A or B)?

  (iii) what is P(A|B)?

(c) If we are given that P(A and B) = 0.1, are the random variables giving rise to events A and B independent?

(d) If we are given that P(A and B) = 0.1, what is P(A|B)?

2.13 Exercise 8 introduces a contingency table summarizing the relationship between hair color and eye color for a group of 1,770 German men. Answer the following questions based on this table.

(a) What is the probability that a randomly chosen man has black hair?

(b) What is the probability that a randomly chosen man has black hair given that he has blue eyes?

(c) What is the probability that a randomly chosen man has black hair given that he has brown eyes?

(d) Are hair color and eye color independent?

2.14 The below contingency table on the type and drive train of cars has been constructed based on the `cars` data set used introduced in Chapter 1. Answer the following questions based on this table.

		Drive Train		
		4WD	front	rear
	large	0	7	4
Type	midsize	0	17	5
	small	2	19	0

(a) What is the probability that a randomly chosen car is midsize?

(b) What is the probability that a randomly chosen car has front wheel drive?

(c) What is the probability that a randomly chosen car has front wheel drive and is midsize?

(d) What is the probability that a randomly chosen car has front wheel drive given that it is midsize?

(e) What is the marginal distribution of drive train?

(f) What is the conditional distribution of drive train for midsize cars, i.e. what are the conditional probabilities is of 4WD, front wheel drive and rear wheel drive when given that the car is midsize?

(g) Are type and drive train independent? Would you be comfortable generalizing your conclusion to all cars?

2.15 Exercise 9 introduces a data set of of 144 cats' body weights and Exercise 10 shows the distribution of gender in this sample of 144 cats. If a randomly chosen cat weighs less than 2.5 *kg*, is it more likely to be male or female?

2.16 Suppose that a student is about to take a multiple choice test that covers algebra and trigonometry. 40% of the questions are algebra and 60% are trigonometry. There is a 90% chance that she will get an algebra question right, and a 35% chance that she will get a trigonometry question wrong.

(a) If we choose a question at random from the exam, what is the probability that she will get it right?

(b) If we choose a question at random from the exam, what is the probability that she will get it wrong?

(c) If we choose a question at random from the exam, what is the probability that it will be an algebra question given that she got it wrong?

2.17 A genetic test is used to determine if people have a genetic predisposition for *thrombosis*, formation of a blood clot inside a blood vessel that obstructs the flow of blood through the circulatory system. It is believed that 3% of the people actually have this predisposition. This test is 99% accurate if a person actually has the predisposition, meaning that the probability of a positive test result when a person actually has the predisposition is 0.99. The test is 98% accurate if a person does not have the predisposition, meaning that the probability of a negative test result when a person does not actually have the predisposition is 0.98. What is the probability that a randomly selected person who is identified as having the predisposition by the test actually has the predisposition?

2.18 Lupus is a medical phenomenon where antibodies that are supposed to attack foreign cells to prevent infections instead see plasma proteins as foreign bodies, leading to a high risk of clotting. It is believed that 2% of the population suffer from this disease.

The test for lupus is very accurate if the person actually has lupus, however is very inaccurate if the person does not. More specifically, the test is 98% accurate if a person actually has the disease, meaning that the probability of a positive test result when a person actually has lupus is 0.98. The test is 74% accurate if a person does not have the disease, meaning that the probability of a negative test result when a person does not actually have lupus is 0.74.

Dr. Gregory House's team presents him a case where the patient tested positive for lupus. However Dr. House claims that "It's never lupus", even though the test result is positive. What do you think? (*Hint: What is the probability that a person who tested positive actually has lupus?*)

2.6.4 Sampling from a small population

2.19 In your sock drawer you have 4 blue socks, 5 grey socks, and 3 black ones. Half asleep one morning you grab 2 socks at random and put them on. Find the probability you end up wearing

(a) 2 blue socks

(b) no grey socks

(c) at least 1 black sock

(d) a green sock

(e) matching socks

2.6.5 Expected value and uncertainty

2.20 You draw a card from a deck. If you get a red card, you win nothing. If you get a spade, you win $5. For any club, you win $10 plus an extra $20 for the ace of clubs.

(a) Create a probability model for the amount you win at this game and find the expected winnings and the standard deviation of the expected winnings.

(b) What is the maximum amount you would be willing to pay to play this game? In other words, what would be the cost of this game to make it fair?

2.21 In the game of roulette, a wheel is spun and you place bets on where it will stop. One popular bet is to bet that it will stop on a red slot. Each slot is equally likely. There are 38 slots, and 18 of them are red. If it does stop on red, you win $1. If not, you lose $1.

(a) Let X represent the amount you win/lose on a single spin. Write a probability model for X.

(b) What's the expected value and standard deviation of your winnings?

2.22 American Airlines charges the following baggage fees: \$20 for the first bag and \$30 for the second. Suppose 54% of passengers have no checked luggage, 34% have one piece of checked luggage and 12% have two pieces. We suppose a negligible portion of people check more than two bags.

(a) Build a probability model and compute the average revenue per passenger.

(b) About how much revenue should they expect for a flight of 120 passengers?

2.23 Andy is always looking for ways to make money fast. Lately he has been trying to make money by gambling. Here is the game he is considering playing:

The game costs \$2 to play. He draws a card from a deck. If he gets a number card (2-10) he wins nothing. For any face card (jack, queen or king) he wins \$3, for any ace he wins \$5 and he wins an extra \$20 if he draws the ace of clubs.

(a) Create a probability model for the amount he wins at this game and find his expected winnings.

(b) Would you recommend this game to Andy as a good way to make money? Explain.

\$	-2	3-2	5-2	20-2
X	2-10	J, Q, K	Ace-3	Ace-♣
P(X)	9/13	3/13	3/52	1/52

$E(X) = (-2)(\frac{9}{13}) + 1(\frac{3}{13}) + 3(\frac{3}{52}) + 18(\frac{1}{52})$
$= -0.63$

Chapter 3

Distributions of random variables

This chapter is a product of OpenIntro and may be printed and shared under a Creative Commons license. To find out about the OpenIntro project or to obtain additional electronic materials or chapters (all free), please go to

openintro.org

3.1 Normal distribution

Among all the distributions we see in practice, one is overwhelmingly the most common. The symmetric, unimodal, bell curve is ubiquitous throughout statistics. Indeed it is so common, that people often know it as the **normal curve** or **normal distribution**[1], shown in Figure 3.1. Variables such as SAT scores and heights of US adult males closely follow the normal distribution.

[1]It is also introduced as the Gaussian distribution after Frederic Gauss, the first person to formalize its mathematical expression.

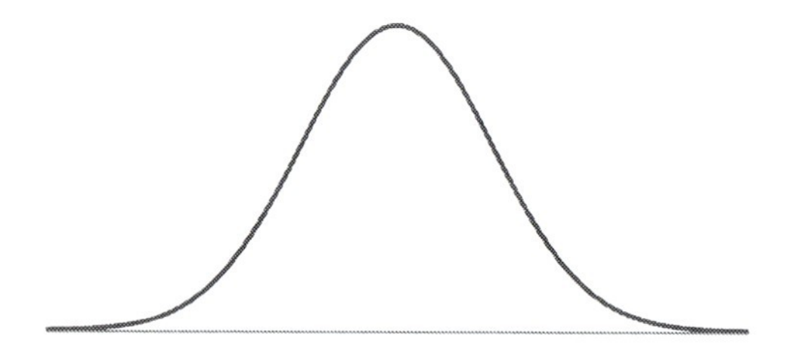

Figure 3.1: A normal curve.

> **Normal distribution facts**
> Many variables are nearly normal, but none are exactly normal. Thus the normal distribution, while not perfect for any single problem, is very useful for a variety of problems. We will use it in data exploration and to solve important problems in statistics.

3.1.1 Normal distribution model

The normal distribution model always describes a symmetric, unimodal, bell shaped curve. However, these curves can look different depending on the details of the model. Specifically, the normal distribution model can be adjusted using two parameters: mean and standard deviation. As you can probably guess, changing the mean shifts the bell curve to the left or right, while changing the standard deviation stretches or constricts the curve. Figure 3.2 shows the normal distribution with mean 0 and standard deviation 1 in the left panel and the normal distributions with mean 19 and standard deviation 4 in the right panel. Figure 3.3 shows these distributions on the same axis.

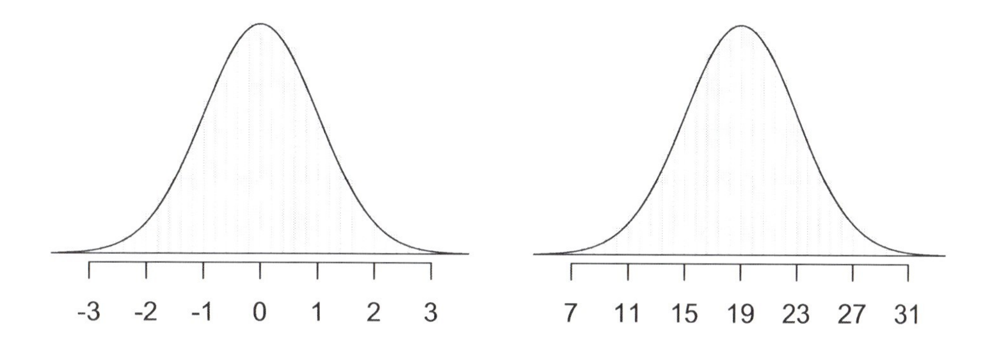

Figure 3.2: Both curves represent the normal distribution, however, they differ in their center and spread.

If a normal distribution has mean μ and standard deviation σ, we may write the distribution as $N(\mu, \sigma)$. The two distributions in Figure 3.2 can be written as $N(0, 1)$ and $N(19, 4)$. Because the mean and standard deviation describe a normal distribution exactly, they are called the distribution's **parameters**.

⊙ **Exercise 3.1** Write down the short-hand for a normal distribution with (a) mean 5 and standard deviation 3, (b) mean -100 and standard deviation 10, and (c) mean 2 and standard deviation 9. The answers for (a) and (b) are in the footnote[2].

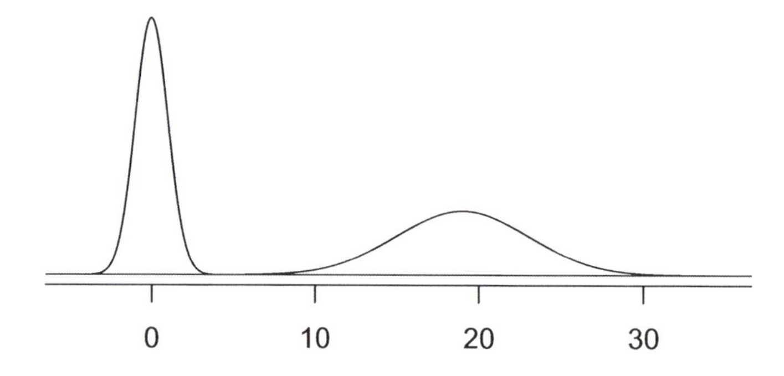

Figure 3.3: The normal models shown in Figure 3.2 but plotted together.

	SAT	ACT
Mean	1500	21
SD	300	5

Table 3.5: Mean and standard deviation for the SAT and ACT.

3.1.2 Standardizing with Z scores

● **Example 3.1** Table 3.5 shows the mean and standard deviation for total scores on the SAT and ACT. The distribution of SAT and ACT scores are both nearly normal. Suppose Pam earned an 1800 on her SAT and Jim obtained a 24 on his ACT. Which of them did better?

We use the standard deviation as a guide. Pam is 1 standard deviation above the average on the SAT: $1500 + 300 = 1800$. Jim is 0.6 standard deviations above the mean: $21 + 0.6 * 5 = 24$. In Figure 3.4, we can see that Pam tends to do better with respect to everyone else than Jim did, so her score was better.

Example 3.1 used a standardization technique called a Z score. The **Z score** of an observation is defined as the number of standard deviations it falls above or below the mean. If it is one standard deviation above the mean, its Z score is 1. If it is 1.5 standard deviations *below* the mean, then its Z score is -1.5. If x is an observation from a distribution[3] $N(\mu, \sigma)$, we mathematically identify the Z score via

$$Z = \frac{x - \mu}{\sigma}$$

Using $\mu_{SAT} = 1500$, $\sigma = 300$, and $x_{Pam} = 1800$, we find Pam's Z score:

$$Z_{Pam} = \frac{x_{Pam} - \mu_{SAT}}{\sigma_{SAT}} = \frac{1800 - 1500}{300} = 1$$

[2] $N(5, 3)$ and $N(-100, 10)$.

[3] It is still reasonable to use a Z score to describe an observation even when x is not nearly normal.

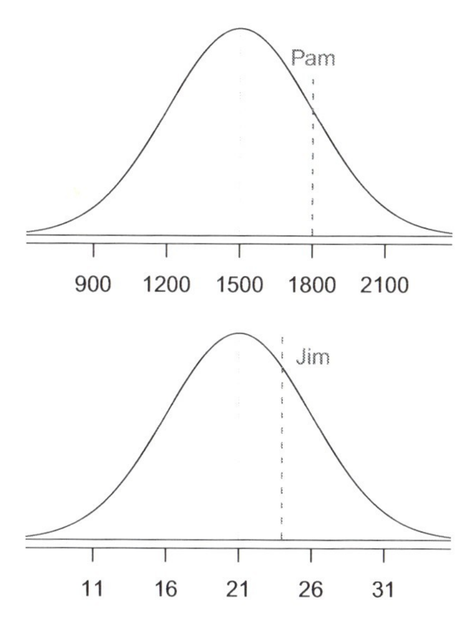

Figure 3.4: Pam's and Jim's scores shown with the distributions of SAT and ACT scores.

The Z score

The Z score of an observation is the number of standard deviations it falls above or below the mean. We compute the Z score for an observation x that follows a distribution with mean μ and standard deviation σ using

$$Z = \frac{x - \mu}{\sigma}$$

⊙ **Exercise 3.2** Use Jim's ACT score, 24, along with the ACT mean and standard deviation to verify his Z score is 0.6.

Observations above the mean always have positive Z scores while those below the mean have negative Z scores. If an observation is equal to the mean (e.g. SAT score of 1500), then the Z score is 0.

⊙ **Exercise 3.3** Let X represent a random variable from $N(3, 2)$, and we observe $x = 5.19$. (a) Find the Z score of x. (b) Use the Z score to determine how many standard deviations above or below the mean x falls. Answers in the footnote[4].

⊙ **Exercise 3.4** The variable `headL` from the `possum` data set is nearly normal with mean 92.6 mm and standard deviation 3.6 mm. Identify the Z scores for $\texttt{headL}_{14} =$

[4](a) Its Z score is given by $Z = \frac{x-\mu}{\sigma} = \frac{5.19-3}{2} = 2.19/2 = 1.095$. (b) The observation x is 1.095 standard deviations *above* the mean. We know it must be above the mean since Z is positive.

95.4 mm and $\texttt{headL}_{79} = 85.8$, which correspond to the 14th and 79th cases in the data set.

We can use Z scores to identify which observations are more unusual than others. One observation x_1 is said to be more unusual than another observation x_2 if the absolute value of its Z score is larger than the absolute value of the other observations Z score: $|Z_1| > |Z_2|$.

⊙ **Exercise 3.5** Which of the observations in Exercise 3.4 is more unusual? [5]

3.1.3 Normal probability table

● **Example 3.2** Pam from Example 3.1 earned a score of 1800 on her SAT with a corresponding $Z = 1$. She would like to know what percentile she falls in for all SAT test-takers.

Pam's **percentile** is the percentage of people who earned a lower SAT score than Pam. We shade this area representing those individuals in Figure 3.6. The total area under the normal curve is always equal to 1, and the proportion of people who scored below Pam on the SAT is equal to the *area* shaded in Figure 3.6: 0.8413. In other words, Pam is in the 84th percentile of SAT takers.

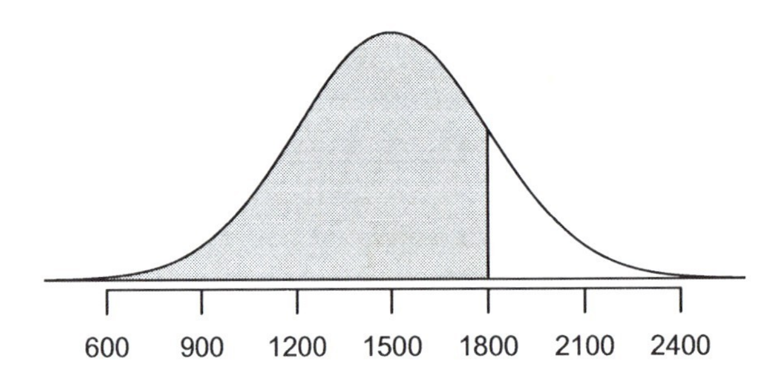

Figure 3.6: The normal model for SAT scores, shading the area of those individuals who scored below Pam.

We can use the normal model to find percentiles. A **normal probability table**, which lists Z scores and corresponding percentiles, can be used to identify a percentile based on the Z score (and vice versa). Statistical software tools, where available, may be used as a substitute for this table.

A normal probability table is given in Appendix A.1 on page 247 and abbreviated in Table 3.8. We use this table to identify the percentile of a Z score. For instance, the percentile of $Z = 0.43$ is shown in row 0.4 and column 0.03 in Table 3.8: 0.6664 or the 66.64^{th} percentile. Generally, we take Z rounded to two decimals, identify the proper row in the normal probability table up through the

[5] In Exercise 3.4, you should have found $Z_{14} = 0.78$ and $Z_{79} = -1.89$. Because the *absolute value* of Z_{79} is larger than Z_{14}, case 79 appears to have a more unusual head length.

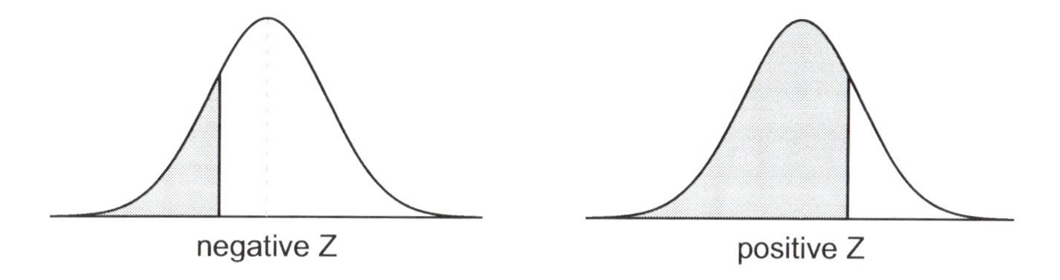

Figure 3.7: The area to the left of Z represents the percentile of the observation.

	Second decimal place of Z									
Z	0.00	0.01	0.02	**0.03**	*0.04*	0.05	0.06	0.07	0.08	0.09
⋮	⋮	⋮	⋮	⋮	⋮	⋮	⋮	⋮	⋮	⋮
0.3	0.6179	0.6217	0.6255	0.6293	0.6331	0.6368	0.6406	0.6443	0.6480	0.6517
0.4	0.6554	0.6591	0.6628	**0.6664**	0.6700	0.6736	0.6772	0.6808	0.6844	0.6879
0.5	0.6915	0.6950	0.6985	0.7019	0.7054	0.7088	0.7123	0.7157	0.7190	0.7224
0.6	0.7257	0.7291	0.7324	0.7357	0.7389	0.7422	0.7454	0.7486	0.7517	0.7549
0.7	0.7580	0.7611	0.7642	0.7673	0.7704	0.7734	0.7764	0.7794	0.7823	0.7852
0.8	0.7881	0.7910	0.7939	0.7967	*0.7995*	0.8023	0.8051	0.8078	0.8106	0.8133
0.9	0.8159	0.8186	0.8212	0.8238	0.8264	0.8289	0.8315	0.8340	0.8365	0.8389
1.0	0.8413	0.8438	0.8461	0.8485	0.8508	0.8531	0.8554	0.8577	0.8599	0.8621
1.1	0.8643	0.8665	0.8686	0.8708	0.8729	0.8749	0.8770	0.8790	0.8810	0.8830
⋮	⋮	⋮	⋮	⋮	⋮	⋮	⋮	⋮	⋮	⋮

Table 3.8: A section of the normal probability table. The percentile for a normal random variable with $Z = 0.43$ has been **highlighted**, and the percentile closest to 0.8000 has also been ***highlighted***.

first decimal, and then determine the column representing the second decimal value. The intersection of this row and column is the percentile of the observation.

We can also find the Z score associated with a percentile. For example, to identify Z for the 80th percentile, we look for the value closest to 0.8000 in the middle portion of the table: 0.7995. We determine the Z score for the 80th percentile by combining the row and column Z values: 0.84.

⊙ **Exercise 3.6** Determine the proportion of SAT test takers who scored better than Pam on the SAT. Hint in the footnote[6].

[6] If 84% had lower scores than Pam, how many had better scores? Generally ties are ignored when the normal model is used.

3.1.4 Normal probability examples

Cumulative SAT scores are approximated well by a normal model, $N(1500, 300)$.

● **Example 3.3** Shannon is a randomly selected SAT taker, and nothing is known about Shannon's SAT aptitude. What is the probability Shannon scores at least 1630 on her SATs?

First, always draw and label a picture of the normal distribution. (Drawings need not be exact to be useful.) We are interested in the chance she scores above 1630, so we shade this upper tail:

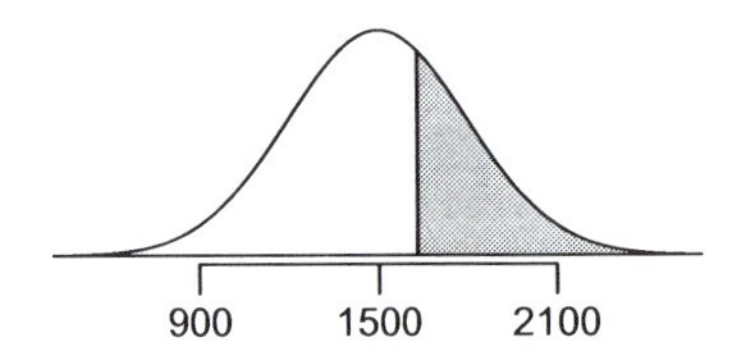

The picture shows the mean and the values at 2 standard deviations above and below the mean. To find areas under the normal curve, we will always need the Z score of the cutoff. With $\mu = 1500$, $\sigma = 300$, and the cutoff value $x = 1630$, the Z score is given by

$$Z = \frac{x - \mu}{\sigma} = \frac{1630 - 1500}{300} = 130/300 = 0.43$$

We look the percentile of $Z = 0.43$ in the normal probability table shown in Table 3.8 or in Appendix A.1 on page 247, which yields 0.6664. However, the percentile describes those who had a Z score *lower* than 0.43. To find the area *above* $Z = 0.43$, we compute one minus the area of the lower tail:

1.0000 − 0.6664 = 0.3336

The probability Shannon scores at least 1630 on the SAT is 0.3336.

TIP: always draw a picture first
For any normal probability situation, *always always always* draw and label the normal curve and shade the area of interest first. The picture will provide a ballpark estimate of the probability.

TIP: find the Z score second
After drawing a figure to represent the situation, identify the Z score for the observation of interest.

⊙ **Exercise 3.7** If the probability of Shannon getting at least 1630 is 0.3336, then what is the probability she gets less than 1630? Draw the normal curve representing this exercise, shading the lower region instead of the upper one. Hint in the footnote[7].

● **Example 3.4** Edward earned a 1400 on his SAT. What is his percentile?

First, a picture is needed. Edward's percentile is the proportion of people who do not get as high as a 1400. These are the scores to the left of 1400.

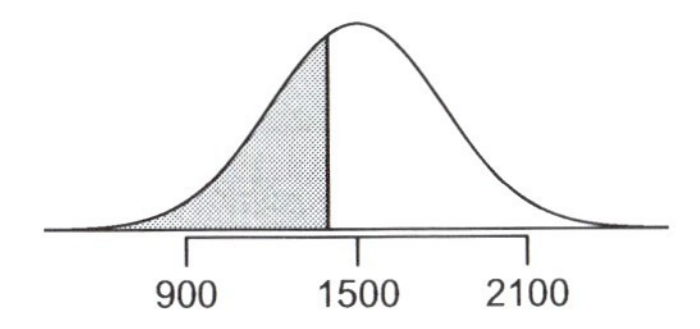

Identifying the mean $\mu = 1500$, the standard deviation $\sigma = 300$, and the cutoff for the tail area $x = 1400$, which makes it easy to compute the Z score:

$$Z = \frac{1400 - \mu}{\sigma} = \frac{1400 - 1500}{300} = -0.33$$

Using the normal probability table, identify the row of -0.3 and column of 0.03, which corresponds to the probability 0.3707. Edward is at the 37th percentile.

⊙ **Exercise 3.8** Use the results of Example 3.4 to compute the proportion of SAT takers who did better than Edward. Also draw a new picture.

TIP: areas to the right
The normal probability table in most books gives the area to the left. If you would like the area to the right, first find the area to the left and then subtract this amount from one.

⊙ **Exercise 3.9** Stuart earned an SAT score of 2100. Draw a picture for each part. (a) What is his percentile? (b) What percent of SAT takers did better than Stuart? Short answers in the footnote[8].

Based on a sample of 100 men, the heights of male adults between the ages 20 and 62 in the US is nearly normal with mean 70.0" and standard deviation 3.3".

⊙ **Exercise 3.10** Mike is 5'7" and Jim is 6'4". (a) What is Mike's height percentile? (b) What is Jim's height percentile? Also draw one picture for each part.

The last several problems have focused on finding the probability or percentile for a particular observation. What if you would like to know the observation corresponding to a particular percentile?

[7]We found the probability in Example 3.3.
[8](a) 0.9772. (b) 0.0228.

● **Example 3.5** Erik's height is at the 40th percentile. How tall is he?

As always, first draw the picture.

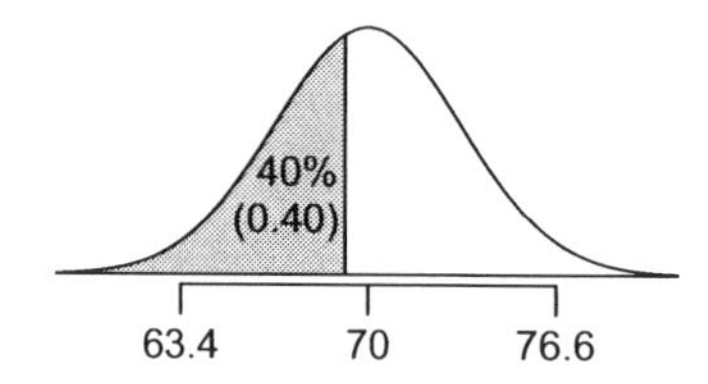

In this case, the lower tail probability is known (0.40), which can be shaded on the diagram. We want to find the observation that corresponds to this value. As a first step in this direction, we determine the Z score associated with the 40th percentile.

Because the percentile is below 50%, we know Z will be negative. Looking in the negative part of the normal probability table, we search for the probability *inside* the table closest to 0.4000. We find that 0.4000 falls in row -0.2 and between columns 0.05 and 0.06. Since it falls closer to 0.05, we take this one: $Z = -0.25$.

Knowing Erik's Z score $Z = -0.25$, $\mu = 70$ inches, and $\sigma = 3.3$ inches, the Z score formula can be setup to determine Erik's unknown height, labeled x:

$$-0.25 = Z = \frac{x - \mu}{\sigma} = \frac{x - 70}{3.3}$$

Solving for x yields the height 69.18 inches[9]. That is, Erik is about 5'9".

● **Example 3.6** What is the adult male height at the 82nd percentile?

Again, we draw the figure first.

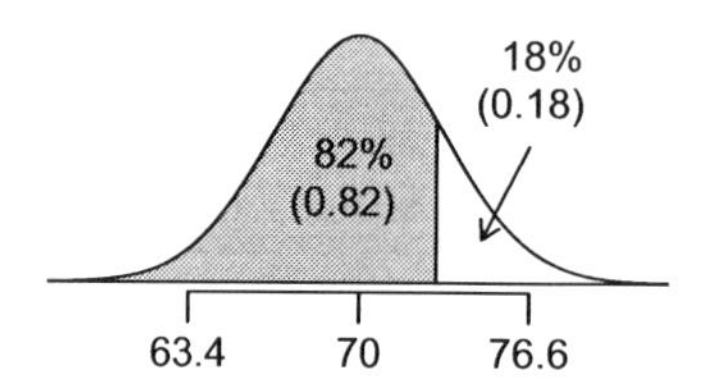

Next, we want to find the Z score at the 82nd percentile, which will be a positive value. Looking in the Z table, we find Z falls in row 0.9 and the nearest column is 0.02, i.e. $Z = 0.92$. Finally, the height x is found using the Z score formula with the known mean μ, standard deviation σ, and Z score $Z = 0.92$:

$$0.92 = Z = \frac{x - \mu}{\sigma} = \frac{x - 70}{3.3}$$

This yields 73.04 inches or about 6'1" as the height at the 82nd percentile.

[9] To solve for x, first multiply by 3.3 and then add 70 to each side.

⊙ **Exercise 3.11** (a) What is the 95th percentile for SAT scores? (b) What is the 97.5^{th} percentile of the male heights? As always with normal probability problems, first draw a picture. Answers in the footnote[10].

⊙ **Exercise 3.12** (a) What is the probability a randomly selected male adult is at least 6'2" (74")? (b) What is the probability a male adult is shorter than 5'9" (69")? Short answers in the footnote[11].

● **Example 3.7** What is the probability a random adult male is between 5'9" and 6'2"?

First, draw the figure. The area of interest is no longer an upper or lower tail.

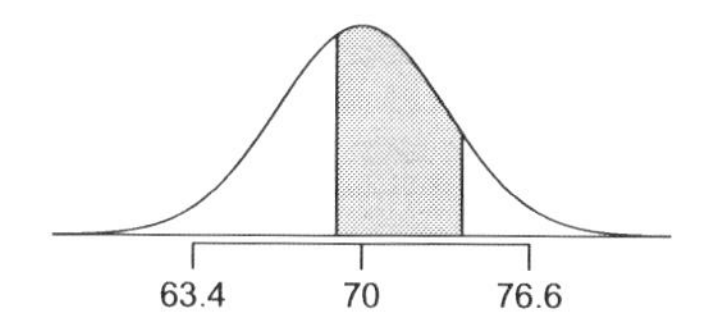

Because the total area under the curve is 1, the area of the two tails that are not shaded can be found (Exercise 3.12): 0.3821 and 0.1131. Then, the middle area is given by

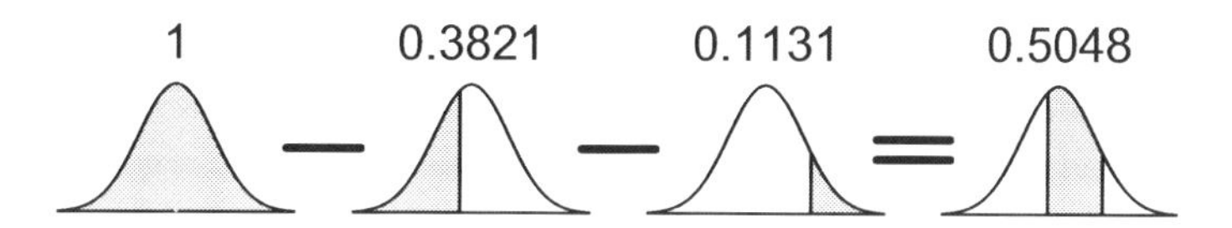

That is, the probability of being between 5'9" and 6'2" is 0.5048.

⊙ **Exercise 3.13** What percent of SAT takers get between 1500 and 2000? Hint in the footnote[12].

⊙ **Exercise 3.14** What percent of adult males are between 5'5" (65") and 5'7" (67")?

3.1.5 68-95-99.7 rule

There are a few rules of thumb all folks should know: the probability of falling within 1, 2, and 3 standard deviations of the mean in the normal distribution. This will be useful in a wide range of practical settings, especially when trying to make a quick estimate without a calculator or Z table.

[10] Remember: draw a picture first, then find the Z score. (We leave the picture to you.) The Z score can be found by using the percentiles and the normal probability table. (a) We look for 0.95 in the probability portion (middle part) of the normal probability table, which leads us to row 1.6 and (about) column 0.05, i.e. $Z_{95} = 1.65$. Knowing $Z_{95} = 1.65$, $\mu = 1500$, and $\sigma = 300$, we setup the Z score formula: $1.65 = \frac{x_{95} - 1500}{300}$. We solve for x_{95}: $x_{95} = 1995$. (b) Similarly, we find $Z_{97.5} = 1.96$, again setup the Z score formula for the heights, and find $x_{97.5} = 76.5$.

[11] (a) 0.1131. (b) 0.3821.

[12] First find the percent who get below 1500 and the percent that get above 2000.

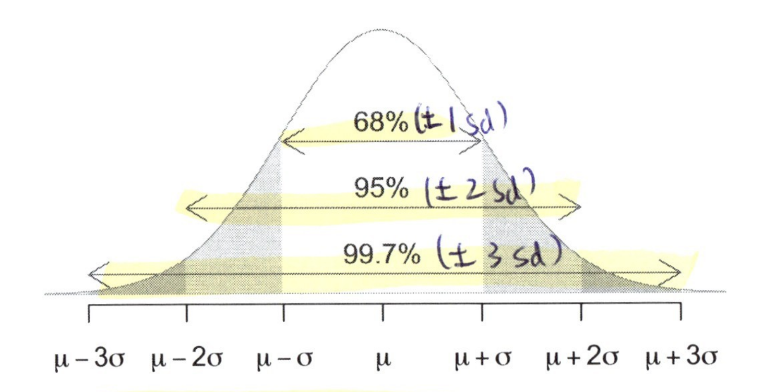

Figure 3.9: Probabilities for falling within 1, 2, and 3 standard deviations for a normal distribution.

⊙ **Exercise 3.15** Confirm that about 68%, 95%, and 99.7% of observations fall within 1, 2, and 3, standard deviations of the mean in the normal distribution, respectively. For instance, first find the area that falls between $Z = -1$ and $Z = 1$, which should have an area of about 0.68. Similarly there should be an area of about 0.95 between $Z = -2$ and $Z = 2$.

It is possible for a normal random variable to fall 4, 5, or even more standard deviations from the mean, however, these occurrences are very rare if the data are nearly normal. The probability of being further than 4 standard deviations from the mean is about 1-in-30,000. For 5 and 6 standard deviations, 1-in-3.5 million and 1-in-1 billion, respectively.

⊙ **Exercise 3.16** SAT scores closely follow the normal model with mean $\mu = 1500$ and standard deviation $\sigma = 300$. (a) About what percent of test takers score 900 to 2100? (b) Can you determine how many score 1500 to 2100? [13].

3.2 Evaluating the normal approximation

Many processes can be well approximated by the normal distribution. We have already seen two good examples: SAT scores and the heights of US adult males. While using a normal model can be extremely convenient and useful, it is important to remember normality is almost always an approximation. Testing the appropriateness of the normal assumption is a key step in practical data analysis.

[13](a) 900 and 2100 represent two standard deviations above and below the mean, which means about 95% of test takers will score between 900 and 2100. (b) Since the normal model is symmetric, then half of the test takers from part (a) ($95\%/2 = 47.5\%$ of all test takers) will score 900 to 1500 while 47.5% score between 1500 and 2100.

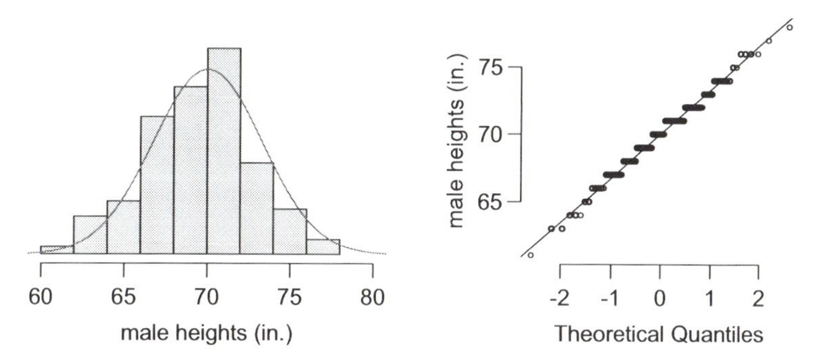

Figure 3.10: A sample of 100 male heights. The observations are rounded to the nearest whole inch, explaining why the points appear to jump in increments in the normal probability plot.

3.2.1 Normal probability plot

Example 3.5 suggests the distribution of heights of US males might be well approximated by the normal model. We are interested in proceeding under the assumption that the data are normally distributed, but first we must check to see if this is reasonable.

There are two visual methods for checking the assumption of normality, which can be implemented and interpreted quickly. The first is a simple histogram with the best fitting normal curve overlaid on the plot, as shown in the left panel of Figure 3.10. The sample mean $\bar{x}$ and standard deviation s are used as the parameters of the best fitting normal curve. The closer this curve fits the histogram, the more reasonable the normal model assumption. Another more common method is examining a **normal probability plot**[14], shown in the right panel of Figure 3.10. The closer the points are to a perfect straight line, the more confident we can be that the data resembles the normal model. We outline the construction of the normal probability plot in Section 3.2.2.

● **Example 3.8** Three data sets of 40, 100, and 400 samples were simulated from a normal probability distribution, and the histograms and normal probability plots of the data sets are shown in Figure 3.11. These will provide a benchmark for what to look for in plots of real data.

The left panels show the histogram (top) and normal probability plot (bottom) for the simulated data set with 40 observations. The data set is too small to really see clear structure in the histogram. The normal probability plot also reflects this, where there are some deviations from the line. However, these deviations are not strong.

The middle panels show diagnostic plots for the data set with 100 simulated observations. The histogram shows more normality and the normal probability plot shows a better fit. While there is one observation that deviates noticeably from the line, it is not particularly extreme.

---

[14]Also commonly called a **quantile-quantile plot**.

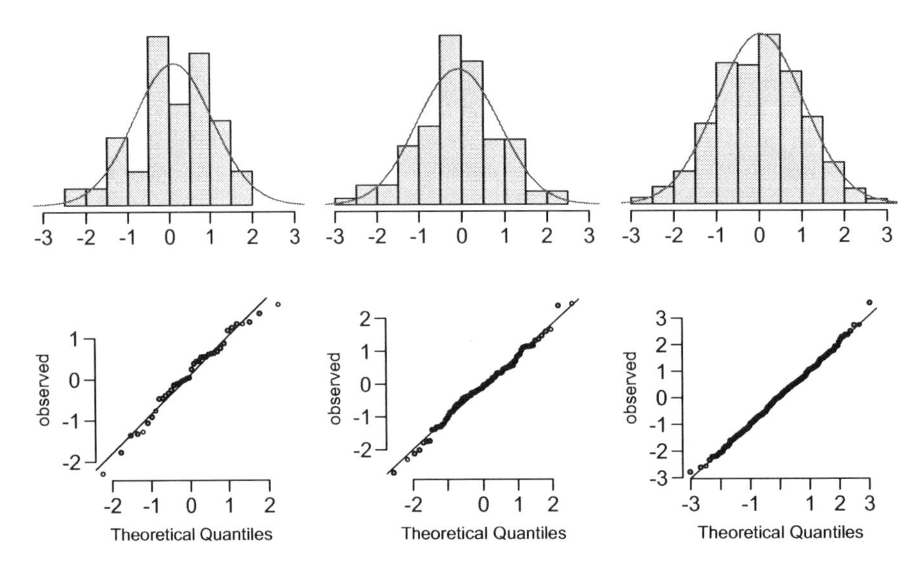

Figure 3.11: Histograms and normal probability plots for three simulated normal data sets; $n = 40$ (left), $n = 100$ (middle), $n = 400$ (right).

The data set with 400 observations has a histogram that greatly resembles the normal distribution while the normal probability plot is nearly a perfect straight line. Again in the normal probability plot there is one observation (the largest) that deviates slightly from the line. If that observation had deviated 3 times further from the line, it would be of much greater concern in a real data set. Apparent outliers can occur in normally distributed data but they are rare and may be grounds to reject the normality assumption.

Notice the histograms look more normal as the sample size increases, and the normal probability plot becomes straighter and more stable. This is generally true when sample size increases.

● **Example 3.9** Are NBA player heights normally distributed? Consider all 435 NBA players from the 2008-9 season presented in Figure 3.12.

We first create a histogram and normal probability plot of the NBA player heights. The histogram in the left panel is slightly left-skewed, which contrasts with the symmetric normal distribution. The points in the normal probability plot do not appear to closely follow a straight line but show what appears to be a "wave". We can compare these characteristics to the sample of 400 normally distributed observations in Example 3.8 and see that they represent much stronger deviations from the normal model. NBA player heights do not appear to come from a normal distribution.

● **Example 3.10** Can we approximate poker winnings by a normal distribution? We consider the poker winnings of an individual over 50 days. A histogram and normal probability plot of this data are shown in Figure 3.13.

The data appear to be strongly right skewed in the histogram, which corresponds to the very strong deviations on the upper right component of the normal probability plot. If we compare these results to the sample of 40 normal observations in

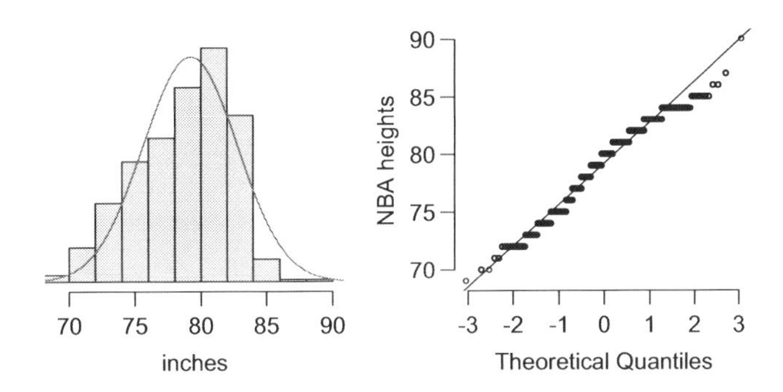

Figure 3.12: Histogram and normal probability plot for the NBA heights from the 2008-9 season.

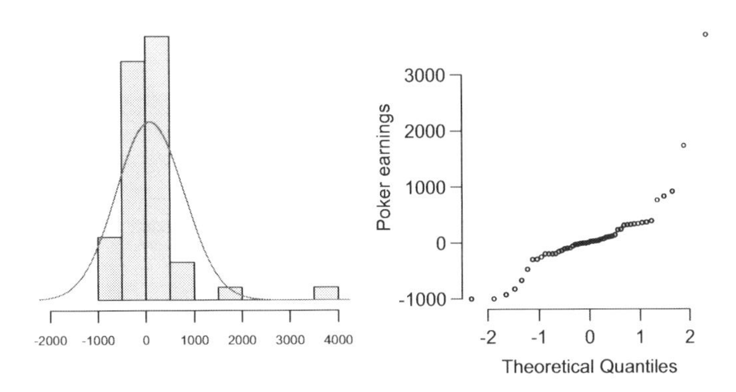

Figure 3.13: A histogram of poker data with the best fitting normal plot and a normal probability plot.

Example 3.8, it is apparent that this data set shows very strong deviations from the normal model. The poker winnings do not appear to follow a normal distribution.

⊙ **Exercise 3.17** Determine which data sets in of the normal probability plots in Figure 3.14 plausibly come from a nearly normal distribution. Are you confident in all of your conclusions? There are 100 (top left), 50 (top right), 500 (bottom left), and 15 points (bottom right) in the four plots. The authors' interpretations are in the footnote[15].

[15] The top-left plot appears show some deviations in the smallest values in the data set; specifically, the left tail of the data set probably has some outliers we should be wary of. The top-right and bottom-left plots do not show any obvious or extreme deviations from the lines for their respective sample sizes, so a normal model would be reasonable for these data sets. The bottom-right plot has a consistent curvature that suggests it is not from the normal distribution. If we examine just the vertical coordinates of these observations, we see that there is a lot of data between -20 and 0, and then about five observations scattered between 0 and 70. This describes a distribution that has a strong right skew.

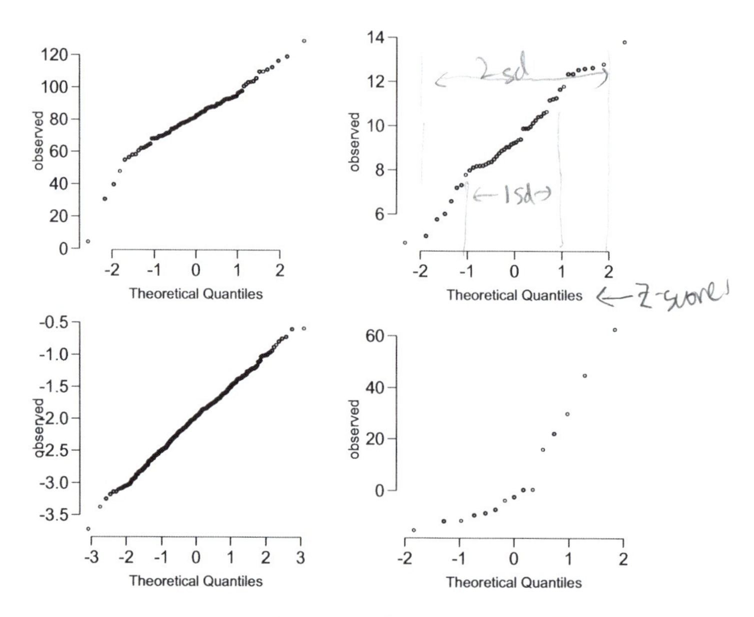

Figure 3.14: Four normal probability plots for Exercise 3.17.

3.2.2 Constructing a normal probability plot*

We construct the plot as follows, which we illustrate in Table 3.15 for the US male height data.

(1) Order the observations.

(2) Determine the percentile of each observation in the ordered data set.

(3) Identify the Z score corresponding to each percentile.

(4) Create a scatterplot of the observations (vertical) against the Z scores (horizontal).

If the observations are normally distributed, then their Z scores will approximately correspond to their percentiles and thus to the z_i in Table 3.15.

Caution: z_i correspond to percentiles
The z_i in Table 3.15 are *not* the Z scores of the observations but only correspond to the percentiles of the observations.

3.3 Geometric distribution

How long should we expect to flip a coin until it turns up `heads`? Or how many times should we expect to roll a die until we get a `1`? These questions can be

Observation i	1	2	3	$\cdots$	100
x_i	61	63	63	$\cdots$	78
Percentile	0.99%	1.98%	2.97%	$\cdots$	99.01%
z_i	-2.33	-2.06	-1.89	$\cdots$	2.33

Table 3.15: Construction of the normal probability plot for the NBA players. The first observation is assumed to be at the 0.99^{th} percentile, and the z_i corresponding to a lower tail of 0.0099 is -2.33. To create the plot based on this table, plot each pair of points, (z_i, x_i).

answered using the geometric distribution. We first formalize each trial – such as a single coin flip or die toss – using the Bernoulli distribution, and then we use these results with our tools from probability (Chapter 2) to construct the geometric distribution.

3.3.1 Bernoulli distribution

Stanley Milgram began a series of experiments in 1963 to estimate what proportion of people would willingly obey an authority and give severe shocks to a stranger. Milgram found that about 65% of people would obey the authority and give such shocks. Over the years, additional research suggested this number is approximately consistent across communities and time[16].

Each person in Milgram's experiment can be thought of as a **trial**. We label a person a **success** if she refuses to administer the worst shock. A person is labeled a **failure** if she administers the worst shock. Because only 35% of individuals refused to administer the most severe shock, we denote the **probability of a success** with $p = 0.35$. The probability of a failure is sometimes denoted with $q = 1 - p$.

Thus, a `success` or `failure` is recorded for each person in the study. When an individual trial only has two possible outcomes, it is called a **Bernoulli random variable**.

Bernoulli random variable, descriptive
A variable with two possible outcomes. We typically label one of these outcomes a "success" and the other outcome a "failure". We may also denote a success by `1` and a failure by `0`.

[16]Find further information on Milgram's experiment at
`http://www.cnr.berkeley.edu/ucce50/ag-labor/7article/article35.htm`.

TIP: "success" need not be something positive
We chose to label a person who refuses to administer the worst shock a "success" and all others as "failures". However, we could just as easily have reversed these labels. The mathematical framework we will build does not depend on which outcome is labeled a success and which a failure.

Bernoulli random variables are often denoted as `1` for a success and `0` for a failure. In addition to being convenient in entering data, it is also mathematically handy. Suppose we observe ten trials:

`0 1 1 1 1 0 1 1 0 0`

Then the **sample proportion**, $\hat{p}$, is the sample mean of these observations:

$$\hat{p} = \frac{\text{\# of successes}}{\text{\# of trials}} = \frac{0+1+1+1+1+0+1+1+0+0}{10} = 0.6$$

This mathematical inquiry of Bernoulli random variables can be extended even further. Because `0` and `1` are numerical outcomes, we can define the mean and standard deviation of a Bernoulli random variable[17].

Bernoulli random variable, mathematical
If X is a random variable that takes value `1` with probability of success p and `0` with probability $1-p$, then X is a Bernoulli random variable with mean and standard deviation

$$\mu = p, \qquad \sigma = \sqrt{p(1-p)}$$

In general, it is useful to think about a Bernoulli random variable as a random process with only two outcomes: a success or failure. Then we build our mathematical framework using the numerical labels `1` and `0` for successes and failures, respectively.

3.3.2 Geometric distribution*

● **Example 3.11** Dr. Smith wants to repeat Milgram's experiments but she only wants to sample people until she finds someone who will not inflict the

[17]If p is the true probability of a success, then the mean of a Bernoulli random variable X is given by

$$\begin{aligned}\mu = E[X] &= P(X=0)*0 + P(X=1)*1 \\ &= (1-p)*0 + p*1 = 0 + p = p\end{aligned}$$

Similarly, the standard deviation of X can be computed:

$$\begin{aligned}\sigma^2 &= P(X=0)(0-p)^2 + P(X=1)(1-p)^2 \\ &= (1-p)p^2 + p(1-p)^2 = p(1-p)\end{aligned}$$

The standard deviation is $\sigma = \sqrt{p(1-p)}$.

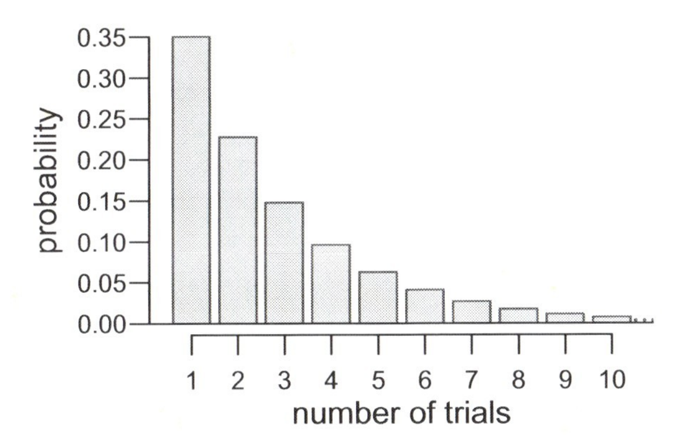

Figure 3.16: The geometric distribution when the probability of success is $p = 0.35$.

worst shock[18]. If the probability a person will *not* give the most severe shock is still 0.35 and the people are independent, what are the chances that she will stop the study after the first person? The second person? The third? What about if it takes her $k-1$ individuals who will administer the worst shock before finding her first success, i.e. the first success is on the kth person? (If the first success is the fifth person, then we say $k = 5$.)

The probability of stopping after the first person is just the chance the first person will not administer the worst shock: $1 - 0.65 = 0.35$. The probability it will be the second person is

$$\begin{aligned} &P(\text{second person is the first to not administer the worst shock}) \\ &= P(\text{the first will, the second won't}) = (0.65)(0.35) = 0.228 \end{aligned}$$

Likewise, the probability it will be the third person is $(0.65)(0.65)(0.35) = 0.148$.

If the first success is on the kth person, then there are $k-1$ failures and then 1 success, which corresponds to the probability $(0.65)^{k-1}(0.35)$. This is the same as $(1-0.35)^{k-1}(0.35)$.

Example 3.11 illustrates what is called the geometric distribution, which describes the waiting time for a success for **independent and identically distributed (iid)** Bernoulli random variables. In this case, the *independence* aspect just means the individuals in the example don't affect each other, and *identical* means they each have the same probability of success.

The geometric distribution from Example 3.11 is shown in Figure 3.16. In general, the probabilities for a geometric distribution decreases **exponentially** fast.

While this text will not derive the formulas for the mean (expected) number of trials needed to find the first success or the standard deviation or variance of this distribution, we present general formulas for each.

[18]This is hypothetical since, in reality, this sort of study probably would not be permitted any longer under current ethical standards.

Geometric Distribution

If the probability of a success in one trial is p and the probability of a failure $1-p$, then the probability of finding the first success in the kth trial is given by

$$(1-p)^{k-1}p \tag{3.1}$$

The mean, i.e. expected value, variance, and standard deviation of this wait time is given by

$$\mu = \frac{1}{p}, \quad \sigma^2 = \frac{1-p}{p^2}, \quad \sigma = \sqrt{\frac{1-p}{p^2}} \tag{3.2}$$

It is no accident that we use the symbol μ for both the mean and expected value. The mean and the expected value are one and the same.

The left side of Equation (3.2) says that, on average, it takes $1/p$ trials to get a success. This mathematical result is consistent with what we would expect intuitively. If the probability of a success is high (e.g. 0.8), then we don't wait for long for a success: $1/0.8 = 1.25$ trials on average. If the probability of a success is low (e.g. 0.1), then we would expect to view many trials before we see a success: $1/0.1 = 10$ trials.

⊙ **Exercise 3.18** The probability an individual would refuse to administer the worst shock is said to be about 0.35. If we were to examine individuals until we found one that did not administer the shock, how many people should we expect to check? The first expression in Equation (3.2) may be useful. The answer is in the footnote[19].

● **Example 3.12** What is the chance that Dr. Smith will find the first success within the first 4 people?

This is the chance it is the first ($k = 1$), second ($k = 2$), third ($k = 3$), or fourth ($k = 4$) person as the first success, which are four disjoint outcomes. Because the individuals in the sample are randomly sampled from a large population, they are independent. We compute the probability of each case and add the separate results:

$$\begin{aligned}
&P(k = 1, 2, 3, \text{ or } 4) \\
&\quad = P(k = 1) + P(k = 2) + P(k = 3) + P(k = 4) \\
&\quad = (0.65)^{1-1}(0.35) + (0.65)^{2-1}(0.35) + (0.65)^{3-1}(0.35) + (0.65)^{4-1}(0.35) \\
&\quad = 0.82
\end{aligned}$$

She has about an 82% chance of ending the study within 4 people.

⊙ **Exercise 3.19** Determine a more clever way to solve Example 3.12. Show that you get the same result. Answer in the footnote[20].

---

[19]We would expect to see about $1/0.35 = 2.86$ individuals to find the first success.

[20]Use the complement: $P(\text{there is no success in the first four observations})$. Compute one minus this probability.

● **Example 3.13** Suppose in one region it was found that the proportion of people who would administer the worst shock was "only" 55%. If people were randomly selected from this region, what is the expected number of people who must be checked before one was found that would be deemed a success? What is the standard deviation of this waiting time?

A success is when someone will **not** inflict the worst shock, which has probability $p = 1 - 0.55 = 0.45$ for this region. The expected number of people to be checked is $1/p = 1/0.45 = 2.22$ and the standard deviation is $\sqrt{(1-p)/p^2} = 1.65$.

⊙ **Exercise 3.20** Using the results from Example 3.13, $\mu = 2.22$ and $\sigma = 1.65$, would it be appropriate to use the normal model to find what proportion of experiments would end in 3 or fewer trials? Answer in the footnote[21].

⊙ **Exercise 3.21** The independence assumption is crucial to the geometric distribution's accurate description of a scenario. Why? Answer in the footnote[22].

3.4 Binomial model*

● **Example 3.14** Suppose we randomly selected four individuals to participate in the "shock" study. What is the chance exactly one of them will be a success? Let's call the four people Allen (A), Brittany (B), Caroline (C), and Damian (D) for convenience. Also, suppose 35% of people are successes as in the previous version of this example.

Let's consider one scenario:

$$\begin{aligned} &P(A = \texttt{refuse},\ B = \texttt{shock},\ C = \texttt{shock},\ D = \texttt{shock}) \\ &\quad = P(A = \texttt{refuse}) * P(B = \texttt{shock}) * P(C = \texttt{shock}) * P(D = \texttt{shock}) \\ &\quad = (0.35) * (0.65) * (0.65) * (0.65) = (0.35)^1(0.65)^3 = 0.096 \end{aligned}$$

But there are three other scenarios: Brittany, Caroline, or Damian could have been the one to refuse. In each of these cases, the probability is again $(0.35)^1(0.65)^3$. These four scenarios exhaust all the possible ways that exactly one of these four people refused to administer the most severe shock so the total probability is $4 * (0.35)^1(0.65)^3 = 0.38$.

⊙ **Exercise 3.22** Verify that the scenario where Brittany is the only one to refuse to give the most severe shock has probability $(0.35)^1(0.65)^3$.

The scenario outlined in Example 3.14 is a special case of what is called the binomial distribution. The **binomial distribution** describes the probability of

[21]No. The geometric distribution is always right skewed and can never be well-approximated by the normal model.

[22]Independence simplified the situation. Mathematically, we can see that to construct the probability of the success on the kth trial, we had to use the Multiplication Rule for Independent processes. It is no simple task to generalize the geometric model for dependent trials.

having exactly k successes in n independent Bernoulli trials with probability of a success p (in Example 3.14, $n = 4$, $k = 1$, $p = 0.35$). We would like to determine the probabilities associated with the binomial distribution more generally, i.e. we want a formula where we can just plug in n, k, and p to obtain the probability. To do this, we reexamine each component of the example.

There were four individuals who could have been the one to refuse, and each of these four scenarios had the same probability. Thus, we could identify the final probability as

$$[\# \text{ of scenarios}] * P(\text{single scenario}) \tag{3.3}$$

The first component of this equation is the number of ways to arrange the $k = 1$ successes among the $n = 4$ trials. The second component is the probability of any of the four (equally probable) scenarios.

Consider P(single scenario) under the general case of k successes and $n - k$ failures in the n trials. In any such scenario, we apply the Product Rule for independent events:

$$p^k(1-p)^{n-k}$$

This is our general formula for P(single scenario).

Secondly, we introduce a general formula for the number of ways to choose k successes in n trials, i.e. arrange k successes and $n - k$ failures:

$$\binom{n}{k} = \frac{n!}{k!(n-k)!}$$

The quantity $\binom{n}{k}$ is called **n choose k** [23]. The exclamation point notation (e.g. $k!$) denotes a **factorial** expression.

$$\begin{aligned}
0! &= 1 \\
1! &= 1 \\
2! &= 2 * 1 = 2 \\
3! &= 3 * 2 * 1 = 6 \\
4! &= 4 * 3 * 2 * 1 = 24 \\
&\vdots \\
n! &= n * (n-1) * ... * 3 * 2 * 1
\end{aligned}$$

Using the formula, we can compute the number of ways to choose $k = 1$ successes in $n = 4$ trials:

$$\binom{4}{1} = \frac{4!}{1!(4-1)!} = \frac{4!}{1!3!} = \frac{4*3*2*1}{(1)(3*2*1)} = 4$$

[23] Other notation for n choose k includes ${}_nC_k$, C_n^k, and $C(n,k)$.

This is exactly what we found by carefully thinking of each possible scenario in Example 3.14.

Substituting n choose k for the number of scenarios and $p^k(1-p)^{n-k}$ for the single scenario probability in Equation (3.3) yields the general binomial formula.

Binomial distribution

Suppose the probability of a single trial being a success is p. Then the probability of observing exactly k successes in n independent trials is given by

$$\binom{n}{k} p^k(1-p)^{n-k} = \frac{n!}{k!(n-k)!}p^k(1-p)^{n-k} \qquad (3.4)$$

Additionally, the mean, variance, and standard deviation of the number of observed successes are

$$\mu = np, \qquad \sigma^2 = np(1-p), \qquad \sigma = \sqrt{np(1-p)} \qquad (3.5)$$

TIP: Is it Binomial? Four conditions to check.
(1) The trials independent.
(2) The number of trials, n, is fixed.
(3) Each trial outcome can be classified as a *success* or *failure*.
(4) The probability of a success (p) is the same for each trial.

- **Example 3.15** What is the probability 3 of 8 randomly selected students will refuse to administer the worst shock, i.e. 5 of 8 will?

We would like to apply the Binomial model, so we check our conditions. The number of trials is fixed ($n = 3$) (condition 2) and each trial outcome can be classified as a success or failure (condition 3). Because the sample is random, the trials are independent and the probability of a success is the same for each trial (conditions 1 and 4).

In the outcome of interest, there are $k = 3$ successes in $n = 8$ trials, and the probability of a success is $p = 0.35$. So the probability that 3 of 8 will refuse is given by

$$\begin{aligned} \binom{8}{3}(0.35)^3(1-0.35)^{8-3} &= \frac{8!}{3!(8-3)!}(0.35)^3(1-0.35)^{8-3} \\ &= \frac{8!}{3!5!}(0.35)^3(0.65)^5 \end{aligned}$$

Dealing with the factorial part:

$$\frac{8!}{3!5!} = \frac{8*7*6*5*4*3*2*1}{(3*2*1)(5*4*3*2*1)} = \frac{8*7*6}{3*2*1} = 56$$

Using $(0.35)^3(0.65)^5 \approx 0.005$, the final probability is about $56 * 0.005 = 0.28$.

TIP: computing binomial probabilities
The first step in using the Binomial model is to check that the model is appropriate. The second step is to identify n, p, and k. The final step is to apply our formulas and interpret the results.

TIP: computing factorials
In general, it is useful to do some cancelation in the factorials immediately. Alternatively, many computer programs and calculators have built in functions to compute n choose k, factorials, and even entire binomial probabilities.

⊙ **Exercise 3.23** If you ran a study and randomly sampled 40 students, how many would you expect to refuse to administer the worst shock? What is the standard deviation of the number of people who would refuse? Equation (3.5) may be useful. Answers in the footnote[24].

⊙ **Exercise 3.24** The probability a random smoker will develop a severe lung condition in his/her lifetime is about 0.3. If you have 4 friends who smoke, are the conditions for the Binomial model satisfied? One possible answer in the footnote[25].

⊙ **Exercise 3.25** Suppose these four friends do not know each other and we can treat them as if they were a random sample from the population. Is the Binomial model appropriate? What is the probability that (a) none of them will develop a severe lung condition? (b) One will develop a severe lung condition? (c) That no more than one will develop a severe lung condition? Answers in the footnote[26].

⊙ **Exercise 3.26** What is the probability that **at least** 2 of your 4 smoking friends will develop a severe lung condition in their lifetimes?

---

[24]We are asked to determine the expected number (the mean) and the standard deviation, both of which can be directly computed from the formulas in Equation (3.5): $\mu = np = 40 * 0.35 = 14$ and $\sigma = \sqrt{np(1-p)} = \sqrt{40 * 0.35 * 0.65} = 3.02$. Because very roughly 95% of observations fall within 2 standard deviations of the mean (see Section 1.3.5, we would probably observe at least 8 but less than 20 individuals in our sample to refuse to administer the shock.

[25]If the friends know each other, then the independence assumption is probably not satisfied.

[26]To check if the Binomial model is appropriate, we must verify the conditions. (i) Since we are supposing we can treat the friends as a random sample, they are independent. (ii) We have a fixed number of trials ($n = 4$). (iii) Each outcome is a success or failure. (iv) The probability of a success is the same for each trials since the individuals are like a random sample ($p = 0.3$ if we say a "success" is someone getting a lung condition, a morbid choice). Compute parts (a) and (b) from the binomial formula in Equation (3.4): $P(0) = \binom{4}{0}(0.3)^0(0.7)^4 = 1 * 1 * 0.7^4 = 0.2401$, $P(1) = \binom{4}{1}(0.3)^1(0.7)^4 = 0.4116$. Note: $0! = 1$, as shown on page 114. Part (c) can be computed as the sum of parts (a) and (b): $P(0) + P(1) = 0.2401 + 0.4116 = 0.6517$. That is, there is about a 65% chance that no more than one of your four smoking friends will develop a severe lung condition.

⊙ **Exercise 3.27** Suppose you have 7 friends who are smokers and they can be treated as a random sample of smokers. (a) How many would you expect to develop a severe lung condition, i.e. what is the mean? (b) What is the probability that at most 2 of your 7 friends will develop a severe lung condition. Hint in the footnote[27].

Below we consider the first term in the Binomial probability, n choose k under some special scenarios.

⊙ **Exercise 3.28** Why is it true that $\binom{n}{0} = 1$ and $\binom{n}{n} = 1$ for any number n? Hint in the footnote[28].

⊙ **Exercise 3.29** How many ways can you arrange one success and $n-1$ failures in n trials? How many ways can you arrange $n-1$ successes and one failure in n trials? Answer in the footnote[29]

[27]First compute the separate probabilities for 0, 1, and 2 friends developing a severe lung condition.

[28]How many different ways are there to arrange 0 successes and n failures in n trials? How many different ways are there to arrange n successes and 0 failures in n trials?

[29]One success and $n-1$ failures: there are exactly n unique places we can put the success, so there are n ways to arrange one success and $n-1$ failures. A similar argument is used for the second question. Mathematically, we show these results by verifying the following two equations:

$$\binom{n}{1} = n, \qquad \binom{n}{n-1} = n$$

3.5 Problem set

3.5.1 Normal distribution

3.1 What percent of a standard normal distribution is found in each region? Be sure to draw a graph.

(a) $Z < -1.35$
(b) $Z > 1.48$
(c) $Z > -1.13$
(d) $Z < 0.18$
(e) $-0.4 < Z < 1.5$
(f) $Z > 8$
(g) $|Z| > 2$
(h) $|Z| < 0.5$

3.2 A college senior who took GRE exam scored 620 on the Verbal Reasoning section and 670 on the Quantitative Reasoning section. The mean score for Verbal Reasoning section was 462 with a standard deviation of 119 and the mean score for the Quantitative Reasoning was 584 with a standard deviation of 151. Assume that both distributions are nearly normal.

(a) Write down the short-hand for these two normal distributions.

(b) What is her Z score on the Verbal Reasoning section? On the Quantitative Reasoning section? Draw a standard normal distribution curve and mark these two Z scores.

(c) What do these Z scores tell you?

(d) Find her percentile scores for the two exams.

(e) On which section did she do better compared to the rest of the exam takers?

(f) What percent of the test takers did better than her on the Verbal Reasoning section? On the Quantitative Reasoning section?

(g) Explain why we cannot just compare her raw scores from the two sections, 620 and 670, and conclude that she did better on the Quantitative Reasoning section.

3.3 Exercise 2 gives the distributions of the scores of the Verbal and Quantitative Reasoning sections of the GRE exam. If the distributions of the scores on these exams cannot be assumed to be nearly normal, how would your answers to parts (b), (c), (d) and (e) of Exercise 2 change?

3.4 Heights of 10 year olds closely follow a normal distribution with mean 55 inches and standard deviation 6 inches.

(a) What is the probability that a randomly chosen 10 year old is shorter than 48 inches?

(b) What is the probability that a randomly chosen 10 year old is between 60 and 65 inches?

(c) If the tallest 10% of the class is considered "very tall", what is the cutoff for "very tall"?

(d) The height requirement for Batman the Ride at Six Flags Magic Mountain is 54 inches. What percent of 10 year olds cannot go on this ride?

3.5 A newspaper article states that the distribution of auto insurance premiums for residents of California is approximately normal with a mean of \$1,650. The article also states that 25% of California residents pay more than \$1,800.

(a) What is the Z score that corresponds to the top 25% (or the 75th percentile) of the standard normal distribution?

(b) In part (a) we got a Z score from the normal probability table. If we wanted to calculate this Z score we would use the formula $Z = \frac{x-\mu}{\sigma}$. Based on the information given in the question what is x and μ?

(c) Use these values to solve for σ, the standard deviation of the distribution.

3.6 The average daily high temperature in June in LA 77 °F with a standard deviation 5 °F. We can assume that the temperatures follow closely a normal distribution.

(a) What is the probability of observing a 82.4 °F temperature or higher in June in LA?

(b) How cold are the coldest 10% of the days in June in LA?

3.7 You feel that the textbook you need to buy for your chemistry class is too expensive at the college book store so you consider buying it on Ebay instead. A look at the past auctions suggest that the prices of the chemistry book you are looking to buy have an approximately normal distribution with mean \$89 and standard deviation \$15.

(a) The textbook costs \$116 at the college book store. Do you think this price is unusually expensive compared to how much people are paying for it on Ebay?

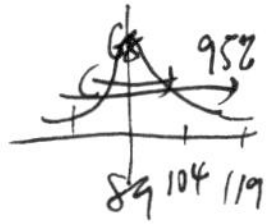

(b) What is the probability that a randomly selected auction for this book closes at more than \$100?

(c) Ebay allows you to set your maximum bid price so that if someone outbids you on an auction you can automatically outbid them, up to the maximum bid price you set. If you are only following one auction, what may be the advantages and disadvantages of setting a too low or too high maximum bid price? What if you are following multiple auctions?

(d) If we watched 10 auctions, what percentile might we use for a maximum bid cutoff to be pretty sure that you will win one of these ten auctions? Why? Is it possible to find a cutoff point that will *ensure* that you win an auction?

(e) If you are patient enough to track ten auctions closely, about what price might you place as your maximum bid price if you want to be pretty sure that you will buy one of these ten books?

3.8 SAT scores (out of 2400) are distributed normally with a mean of 1500 and a standard deviation of 300. What percentage of students at this university scored below 2100, given that they needed to have scored 1900 to get in?

3.9 Below are final exam scores of 20 Introductory Statistics students. Also provided are some sample statistics. Use this information to determine if the scores follow approximately the 68-95-99.7% Rule.

79, 83, 57, 82, 94,
83, 72, 74, 73, 71,
66, 89, 78, 81, 78,
81, 88, 69, 77, 79

Mean	77.7
Variance	71.27
Std. Dev.	8.44

3.5.2 Evaluating the Normal distribution

3.10 Exercise 9 lists the final exam scores of 20 Introductory Statistics students. Do these data follow a normal distribution? Explain your reasoning using the graphs provided below.

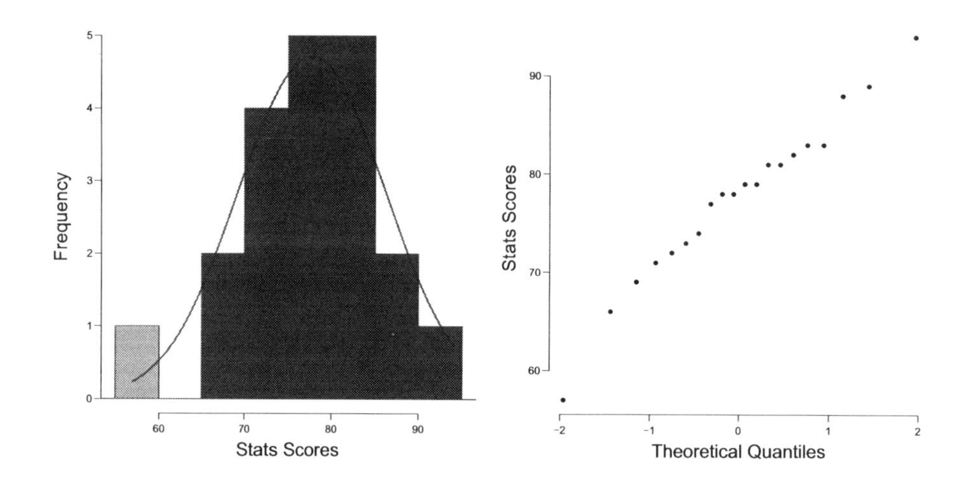

3.5.3 Geometric distribution

3.11 In a hand of poker, can each card dealt be considered an independent Bernoulli trial?

3.12 You have a fair die.

(a) What is the probability of not rolling a 6 on the 1st try and rolling a 6 on the 2nd try?

(b) How many times would you expect to roll before getting a 6? What is the standard deviation?

3.13 You have an unfair die such that the probability of rolling a six is 1/10.

(a) What is the probability of not rolling a 6 on the 1st try and rolling a 6 on the 2nd try?

(b) How many times would you expect to roll before getting a 6? What is the standard deviation?

3.14 Based on your answers to exercises 12 and 13, how does decreasing the probability of an event affect the mean and standard deviation of the wait time until success?

3.15 The probability that two brown eyed parents will have a brown eyed child is 0.75. There is also a 0.125 probability that the child will have blue eyes and a 0.125 probability that the child will have green eyes. [7]

(a) What is the probability that the first two children of two brown eyed parents do not have blue eyes and the third child does? Assume that the eye colors of the children are independent of each other.

(b) How many children should they expect to have until the first blue eyed child? What is the standard deviation?

3.5.4 Binomial distribution

3.16 According to a 2008 study, 69.7% of 18-20 year olds consumed alcoholic beverages in the past year. [8]

(a) Can we use the binomial distribution to calculate the probability of finding exactly six people out of a random sample of ten 18-20 year olds who consumed alcoholic beverages? Explain.

(b) Calculate this probability.

(c) What is the probability that exactly four out of ten 18-20 year olds have *not* consumed an alcoholic beverage?

3.17 Based on the information given in Exercise 16, in a group of fifty 18-20 year olds,

(a) how many people would you expect to have consumed alcoholic beverages? And with what standard deviation?

(b) would you be surprised if there were 45 or more people who have consumed alcoholic beverages?

3.18 Based on the information given in Exercise 16, in a group of five 18-20 year olds, what is the probability that

(a) at most 2 of them have consumed alcoholic beverages?

(b) at least 1 of them have consumed alcoholic beverages?

3.19 If you flip a fair coin 10 times, what is the probability of

(a) getting at least one tail?

(b) getting exactly 5 tails?

(c) getting exactly 5 tails and 5 heads?

(d) getting exactly 3 tails and 4 heads?

3.20 Based on the information given on Exercise 15, if two brown eyed parents

(a) have two children, what is the probability that

 (i) their first child will have green eyes and the second will not?

 (ii) exactly one will have green eyes?

(b) have six children, what is the probability that

 (i) two will have green eyes?

 (ii) none will have green eyes?

 (iii) at least one will have green eyes?

 (iv) the first green eyed child will be the 4th child?

3.21 In the game of roulette, a wheel is spun and you place bets on where it will stop. One popular bet is to bet that it will stop on a red slot. Each slot is equally likely. There are 38 slots, and 18 of them are red. If it does stop on red, you win $1. If not, you lose $1.

Suppose you play 3 times. Let Y represent the total amount won or lost. Write a probability model for Y.

3.22 In a multiple choice exam there are 5 questions and 4 choices for each question (a, b, c, d). Robin has not studied for the exam at all, and decided to randomly guess the answers. What is the probability that

(a) the first question she gets right is the 3rd question?

(b) she gets exactly 3 or exactly 4 questions right?

(c) she gets majority of the questions right?

Chapter 4

Foundations for inference

Statistical inference is concerned primarily with understanding features and the quality of parameter estimates. For example, a classic inferential question is "How sure are we that the true population mean, μ, is near our estimated mean, $\bar{x}$?" While the equations and details change depending on the parameter we are studying[1], the foundations for inference are the same throughout all of statistics. We introduce these common themes in Sections 4.1-4.4 by discussing inference about the population mean, μ, and set the stage for other parameters in Section 4.5. Understanding this Chapter well will make the rest of the book, and indeed the rest of Statistics, seem much more familiar.

Throughout the next few sections we consider a data set called `run10`, shown in Table 4.1. The `run10` data set represents all 14,974 runners who finished the the 2009 Cherry Blossom 10 mile run in Washington, DC. The variables are described in Table 4.2.

This data is special because it includes the results for the entire population of runners who finished the 2009 Cherry Blossom Run. We take a simple random sample of this population, `run10Samp`, represented in Table 4.3. Using this sample, we will draw conclusions about the entire population. This is the practice of statistical inference in the broadest sense.

[1] We have already seen μ, σ, and p, and we will be introduced to others during later chapters.

	time	age	gender	state
1	94.12	32	M	MD
2	71.53	33	M	MD
⋮	⋮	⋮	⋮	⋮
14974	83.43	29	F	DC

Table 4.1: Three observations from the `run10` data set.

variable	**description**
`time`	ten mile run time, in minutes
`age`	age, in years
`gender`	gender (`M` for male, `F` for female, and `N` for not-listed)
`state`	home state (or country if not from US)

Table 4.2: Variables and their descriptions for the `run10` data set.

	time	age	gender	state
1	96.20	46	F	VA
2	69.43	20	M	MD
⋮	⋮	⋮	⋮	⋮
100	78.73	61	F	NY

Table 4.3: Three observations for the `run10Samp` data set, which represents a random sample of 100 runners from the 2009 Cherry Blossom Race.

4.1 Variability in estimates

We would like to estimate two features of the Cherry Blossom runners using the `run10Samp` sample.

(1) How long does it take a runner, on average, to complete the 10 miles?

(2) What is the average age of the runners?

These questions are important when considering planning future events[2]. We will use $x_1, ..., x_{100}$ to represent the ten mile time for each runner in our sample, and $y_1, ..., y_{100}$ will represent the age of each of these participants.

4.1.1 Point estimates

We want to estimate the **population mean** based on the sample. The most intuitive way to go about doing this is to simply take the **sample mean**. That is, to estimate the average 10 mile run time of all participants, take the average time for the sample:

$$\bar{x} = \frac{96.2 + 69.43 + \cdots + 78.73}{100} = 93.65$$

The sample mean $\bar{x} = 93.65$ minutes is called a **point estimate** of the population mean: if we can only choose one value to estimate the population mean, this is our best guess. Will this point estimate be exactly equal to the population mean? No, but it will probably be close.

⊙ **Exercise 4.1** Why won't the point estimate be exactly equal to the population mean? Answer in the footnote[3].

Likewise, we can estimate the average age of participants by taking the sample mean of `age`:

$$\bar{y} = \frac{46 + 20 + \cdots + 61}{100} = 35.22$$

The average age of the participants in the sample is 35.22 years.

What about generating point estimates of other **population parameters**, such as the population median or population standard deviation? Here again we might estimate parameters based on sample statistics, as shown in Table 4.4. For example, we estimate the population standard deviation for the running time using the sample standard deviation, 15.66 minutes.

[2]While we focus on the mean in this chapter, questions regarding variation are many times equally important in practice. For instance, we would plan an event very differently if the standard deviation of runner age was 2 versus if it was 20.

[3]If we take a new sample of 100 people and recompute the mean, we probably will not get the exact same answer. Our estimate will vary from one sample to another. This *sampling variation* suggests our estimate may be close but not exactly equal to the parameter.

time	estimate	parameter
mean	93.65	94.26
median	92.51	93.85
st. dev.	15.66	15.44

Table 4.4: Point estimates and parameters for the `time` variable.

⊙ **Exercise 4.2** What is the point estimate for the standard deviation of the amount of time it takes participants to run 10 miles? Table 4.4 may be helpful.

⊙ **Exercise 4.3** Suppose we want to estimate the difference in run times for men and women. If $\bar{x}_{men} = 90.10$ and $\bar{x}_{women} = 96.92$, then what would be a good point estimate for the population difference? Answer in the footnote[4].

⊙ **Exercise 4.4** If you had to provide a point estimate of the population IQR for the run time of participants, how might you make such an estimate using a sample? Answer in the footnote[5].

4.1.2 Point estimates are not exact

Estimates are usually not exactly equal to the truth, but they get better as more data become available. We can see this by plotting a **running mean** from our `run10Samp` sample. A **running mean** is a sequence of means, where each mean uses one more observation in its calculation than the mean directly before it in the sequence. For example, the second mean in the sequence is the average of the first two observations and the third in the sequence is the average of the first three. The running mean is shown for the `time` variable in Figure 4.5, and it approaches the true population average, 94.26 minutes, as more data becomes available.

Sample point estimates only approximate the population parameter, and they vary from one sample to another. If we took another simple random sample of the Cherry Blossom runners, we would find that the sample mean for `time` would be a little different. It will be useful to quantify how variable an estimate is from one sample to another. If this variability is small (i.e. the sample mean doesn't change much from one sample to another) then that estimate is probably very accurate. If it varies widely from one sample to another, then we should not expect our estimate to be very good.

[4]We could take the difference of the two sample means: $96.92 - 90.10 = 6.82$. Men appear to have run about 6.82 minutes faster on average in the 2009 Cherry Blossom Run.

[5]To obtain a point estimate of the IQR for the population, we could take the IQR of the sample.

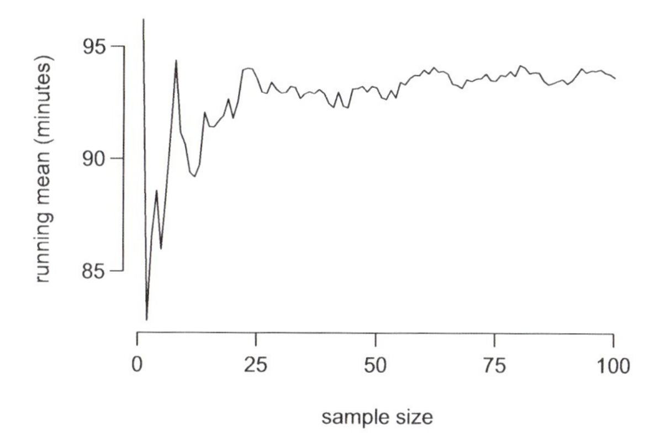

Figure 4.5: The mean computed after adding each individual to the sample. The mean tends to approach the true population average as more data become available.

4.1.3 Standard error of the mean

From the random sample represented in `run10Samp`, we guessed the average time it takes to run 10 miles is 93.65 minutes. Suppose we take another random sample of 100 individuals and take its mean: 95.05 minutes. Suppose we took another (94.35 minutes) and another (93.20 minutes) and so on. If we do this many many times – which we can do only because we have the entire population data set – we can build up a **sampling distribution** for the sample mean when the sample size is 100, shown in Figure 4.6.

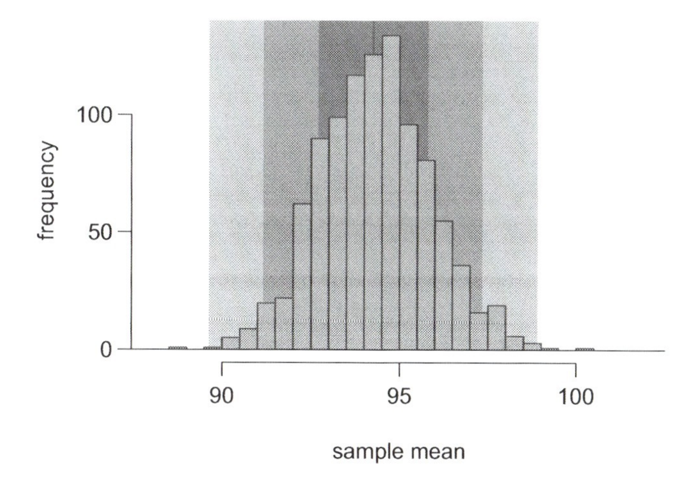

Figure 4.6: A histogram of the sample means for 1000 different random samples of size $n = 100$.

Sampling distribution
The sampling distribution represents the distribution of the point estimate. It is useful to think of the point estimate as being plucked from such a distribution. Understanding the concept of a sampling distribution is central to understanding statistical inference.

The sampling distribution shown in Figure 4.6 can be described as follows. The distribution is approximately symmetric, unimodal, and is centered at 94.26 minutes. The variability of the point estimates is described by the standard deviation of this distribution, which is approximately 1.5 minutes.

The center of the distribution, 94.26 minutes, is the actual population mean μ. Intuitively, this makes sense. The sample means should tend to be close to and "fall around" the population mean.

We can see that the sample mean has some variability around the population mean, which we can quantify using the standard deviation of the sample mean: $\sigma_{\bar{x}} = 1.54$. In this case, the standard deviation tells us how far the typical estimate is away from the actual population mean, 94.26 minutes. Thus, this standard deviation describes the typical **error**, and for this reason we usually call this standard deviation the **standard error** of the estimate.

Standard error of an estimate
The standard deviation associated with an estimate is called the *standard error*. It describes the typical error or uncertainty associated with the estimate.

When considering the case of the point estimate $\bar{x}$, there is one problem: there is no obvious way to estimate its standard error from a single sample. However, there is a way around this issue.

⊙ **Exercise 4.5** (a) Would you rather use a small sample or a large sample when estimating a parameter? Why? (b) Using your reasoning from (a), would you expect a point estimate based on a small sample to have smaller or larger standard error than a point estimate based on a larger sample? Answers in the footnote[6].

In the sample of 100 observations, the standard error of the sample mean is equal to one-tenth of the population standard deviation: $1.54 = 15.4/10$. In other words, the standard error of the sample mean based on 100 observations is equal

[6](a) Consider two random samples: one of size 10 and one of size 1000. The larger sample would tend to provide a more accurate estimate. Individual observations in the small sample are highly influential on the estimate while in larger samples these individual observations would more often average each other out. (b) If we think an estimate is better, we probably mean it typically has less error. Based on (a), our intuition suggests that the larger the sample size the smaller the standard error.

to

$$SE_{\bar{x}} = \sigma_{\bar{x}} = \frac{\sigma_x}{\sqrt{100}} = \frac{15.44}{\sqrt{100}} = 1.54$$

where σ_x is the standard deviation of the individual observations. This is no coincidence. We can show mathematically that this equation is correct when the observations are independent[7]. The most common way to verify observations in a sample are independent is to collect a simple random sample of less than 10% of the population[8].

Computing SE for the sample mean

Given n independent observations from a population with standard deviation σ, the standard error of the sample mean is equal to

$$SE = \frac{\sigma}{\sqrt{n}} \tag{4.1}$$

A reliable method to ensure sample observations are independent is to conduct a simple random sample consisting of less than 10% of the population.

There is one subtle issue of Equation (4.1): the population standard deviation is typically unknown. But you might have already guessed how to resolve this problem: we use the point estimate of the standard deviation from the sample. This estimate tends to be sufficiently good when the sample size is at least 50 and not extremely skewed. Thus, we often just use the sample standard deviation s instead of σ. When the sample size is smaller than 50 or the data is strongly skewed, we must beef up our tools to compensate, which is the topic of Chapter 6.

⊙ **Exercise 4.6** In Section 4.1.1, the sample standard deviation of the runner age for `run10Samp` was $s_y = 10.93$ based on the sample of size 100. Because the sample is simple random and consists of less than 10% of the population, the observations are independent. (a) Compute the standard error of the sample mean, $\bar{y} = 35.22$ years. (b) Would you be surprised if someone told you the average age of all the runners was actually 35 years? Answers in the footnote[9].

⊙ **Exercise 4.7** (a) Would you be more trusting of a sample that has 100 observations or 400 observations? (b) We want to show mathematically that our estimate tends to

[7]This can be shown based on the probability tools of Section 2.6. [Authors' note: Section 2.6 will be added for the First Edition released in 2011.]

[8]The choice of 10% is based on the findings of Section 2.4; observations look like they were independently drawn so long as the sample consists of less than 10% of the population.

[9](a) Because observations are independent, use Equation (4.1) to compute the standard error: $SE_{\bar{y}} = 10.93/\sqrt{100} = 1.09$ years. (b) It would not be surprising. Our sample is within about 0.2 standard errors of 35 years. In other words, 35 years old does not seem to be implausible given that our sample was so close to it. (We use our standard error to identify what is close. 0.2 standard errors is very close.)

be better when the sample size is larger. If the standard deviation of the individual observations is 10, what is our estimate of the standard error when the sample size is 100? What about when it is 400? (c) Explain how your answer to (b) mathematically justifies your intuition in part (a). Answers are in the footnote[10].

4.1.4 Summary

In this section, we have achieved three goals. First, we have determined that we can use point estimates from a sample to estimate population parameters. We have also determined that these point estimates are not exact: they are very from one sample to another. Lastly, we have quantified the uncertainty of the sample mean using what we call the standard error, mathematically represented in Equation (4.1). While we could also quantify the standard error for other estimates – such as the median, standard deviation, or any other number of statistics – we postpone this generalization for later chapters or courses.

4.2 Confidence intervals

A point estimate provides a single plausible value for a parameter. However, a point estimate is rarely perfect; usually there is some error in the estimate. Instead of supplying just a point estimate of a parameter, a next logical step would be to provide a plausible *range of values* for the parameter.

In this section and in Section 4.3, we will emphasize the special case where the point estimate is a sample mean and the parameter is the population mean. In Section 4.5, we will discuss how we generalize the methods to apply to a variety of point estimates and population parameters that we will encounter in Chapter 5.

4.2.1 Capturing the population parameter

A plausible range of values for the population parameter is called a **confidence interval**.

Using only a point estimate is like fishing in a murky lake with a spear, and using a confidence interval is like fishing with a net. We can throw a spear where we saw a fish but we will probably miss. On the other hand if we toss a net in that area, we have a good chance of catching the fish.

If we report a point estimate, we probably will not hit the exact population parameter. On the other hand, if we report a range of plausible values – a confidence interval – we have a good shot at capturing the parameter.

[10](a) Larger samples tend to have more information, so a point estimate with 400 observations seems more trustworthy. (b) The standard error when the sample size is 100 is given by $SE_{100} = 10/\sqrt{100} = 1$. For 400: $SE_{400} = 10/\sqrt{400} = 0.5$. The larger sample has a smaller standard error. (c) The (standard) error of the sample with 400 observations is lower than the sample with 100 observations. Since the error is lower, this mathematically shows the estimate from the larger sample tends to be better (though it does not guarantee that a particular small sample might just happen to provide a better estimate than a particular large sample).

⊙ **Exercise 4.8** If we want to be very certain we capture the population parameter, should we use a wider interval or a smaller interval? Answer in the footnote[11].

4.2.2 An approximate 95% confidence interval

Our point estimate is the most plausible value of the parameter, so it makes sense to cast the confidence interval around the point estimate. The standard error, which is a measure of the uncertainty associated with the point estimate, provides a guide for how large we should make the confidence interval.

The standard error represents the standard deviation associated with the estimate, and roughly 95% of the time the estimate will be within 2 standard errors of the parameter. If the interval spreads out 2 standard errors from the point estimate, we can be roughly 95% **confident** we have captured the true parameter:

$$\text{point estimate} \pm 2 * SE \tag{4.2}$$

But what does "95% confident" mean? Suppose we took many samples and built a confidence interval for the mean for each sample using Equation (4.2). Then about 95% of those intervals would contain the actual mean, μ. We have done this in Figure 4.7 with 25 samples, where 24 of the resulting confidence intervals contain the average time for all the runners, $\mu = 94.26$ minutes, and one does not.

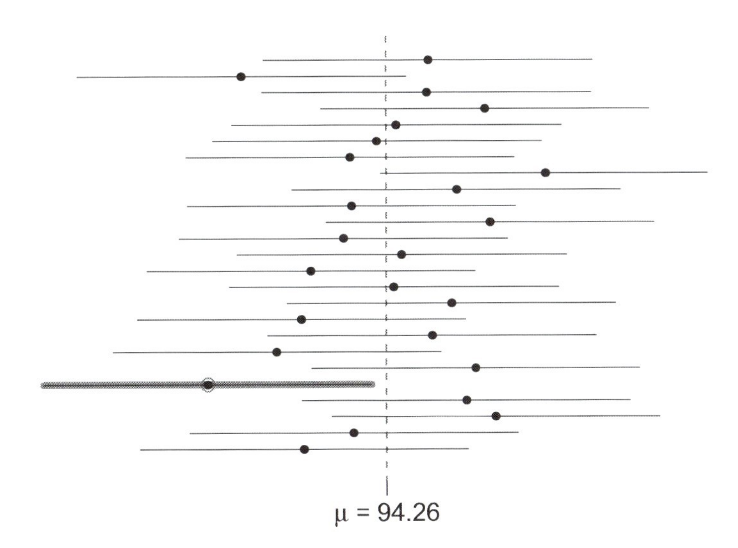

Figure 4.7: Twenty-five samples of size $n = 100$ were taken from the `run10` data set. For each sample, a confidence interval was created to try to capture the average 10 mile time. Only 1 of the 25 intervals does not capture the true mean, $\mu = 94.26$ minutes.

[11] If we want to be more certain we will capture the fish, we might use a wider net. Likewise, we use a wider interval if we want to be more certain we capture the parameter.

The rule where about 95% of observations are within 2 standard deviations of the mean is only approximately true. However, it holds very well for the normal distribution. As we will soon see, the mean tends to be normally distributed when the sample size is sufficiently large.

● **Example 4.1** If the sample mean of the 10 mile run times from `run10Samp` is 93.65 minutes and the standard error is 1.57 minutes, what would be an approximate 95% confidence interval for the average 10 mile time for all runners in the race?

We apply Equation (4.2):

$$93.65 \pm 2 * 1.57 \quad \rightarrow \quad (90.51, 96.79)$$

Based on this data, we are about 95% confident that the average 10 mile time for all runners in the race was larger than 90.51 but less than 96.76 minutes. Our interval extends out 2 standard errors from the point estimate, $\bar{x}$.

⊙ **Exercise 4.9** The `run10Samp` sample suggested the average runner age is about 35.22 years with a standard error of 1.09 years. What is a 95% confidence interval for the average age of all of the runners? Answer in the footnote[12].

⊙ **Exercise 4.10** In Figure 4.7, one interval does not contain 94.26 minutes. Does this imply that the mean cannot be 94.26? Answer in the footnote[13].

4.2.3 A sampling distribution for the mean

In Section 4.1.3, we introduced a sampling distribution for $\bar{x}$, the average run time for samples of size 100. Here we examine this distribution more closely by taking samples from the `run10` data set, computing the mean of `time` for each sample, and plotting those sample means in a histogram. We did this earlier in Figure 4.6, however, we now take 10,000 samples to get an especially accurate depiction of the sample distribution. A histogram of the means from the 10,000 samples is shown in the left panel of Figure 4.8.

Does this distribution look familiar? Hopefully so! The distribution of sample means closely resembles the normal distribution (see Section 3.1). A normal probability plot of these sample means is shown in the right panel of Figure 4.8. Because all of the points closely fall around a straight line, we can conclude the distribution of sample means is nearly normal (see Section 3.2). This result can be explained by the Central Limit Theorem.

[12] Again apply Equation (4.2): $35.22 \pm 2 * 1.09 \rightarrow (33.04, 37.40)$. We interpret this interval as follows: We are about 95% confident the average age of all participants in the 2009 Cherry Blossom Run was between 33.04 and 37.40 years.

[13] Just as some observations occur more than 2 standard deviations from the mean, some point estimates will be more than 2 standard errors from the parameter. A confidence interval only provides a plausible range of values for a parameter. While we might say other values are implausible, this does not mean they are impossible.

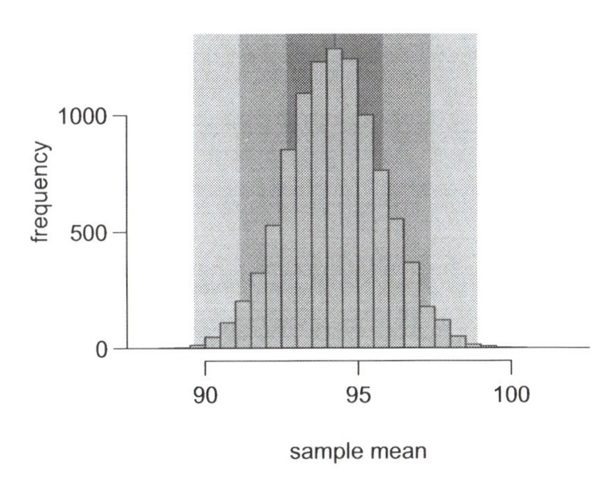

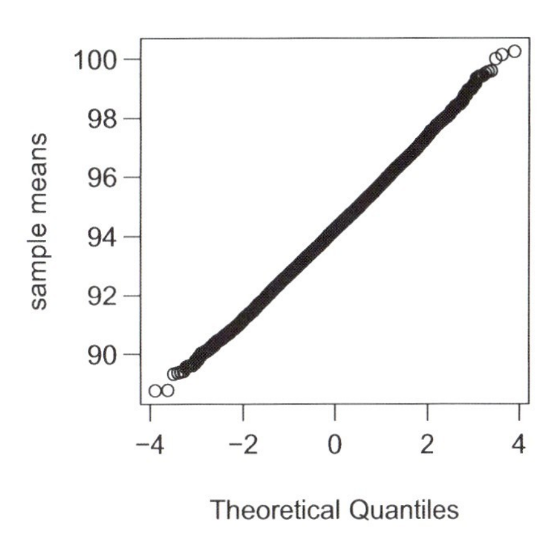

Figure 4.8: The left panel shows a histogram of the sample means for 10,000 different random samples. The right panel shows a normal probability plot of those sample means.

> **Central Limit Theorem, informal description**
> If a sample consists of at least 50 independent observations and the data is not extremely skewed, then the distribution of the sample mean is well approximated by a normal model.

We revisit the Central Limit Theorem in more detail in Section 4.4. However, we will apply this informal rule for the time being.

The choice of using 2 standard errors in Equation (4.2) was based on our general guideline that roughly 95% of the time, observations are within two standard deviations of the mean. Under the normal model, we can actually make this more accurate by using 1.96 in place of 2.

> **95% Confidence interval**
> If a point estimate such as $\bar{x}$ is associated with a normal model and standard error SE, then a 95% confidence interval for the parameter is
>
> $$\text{point estimate} \pm 1.96 * SE \tag{4.3}$$

> **Margin of error**
> In a 95% confidence interval, $1.96 * SE$ is called the **margin of error**.

4.2.4 Changing the confidence level

Suppose we want to consider confidence intervals where the confidence level is somewhat higher than 95%: perhaps we would like a confidence level of 99%. Thinking back to trying to catch the fish, if we want to be even more sure we will catch the fish, we use a wider net. To create a 99% confidence level, we must also widen our 95% interval. On the other hand, if we want an interval with lower confidence, such as 90%, we could make our interval slightly slimmer than a 95% interval.

We can break down the 95% confidence interval structure to determine how to make intervals with new confidence levels. Below is a 95% confidence interval:

$$\text{point estimate} \pm 1.96 * SE \tag{4.4}$$

There are three components to this interval: the point estimate, "1.96", and the standard error. The choice of $1.96 * SE$ was based on capturing 95% of the data since the estimate is within *1.96* standard deviations of the parameter about 95% of the time. The choice of 1.96 corresponds to a 95% confidence level.

⊙ **Exercise 4.11** If X is a normally distributed random variable, how often will X be within 2.58 standard deviations of the mean? Answer in the footnote[14].

To create a 99% confidence interval, adjust 1.96 in the 95% confidence interval formula to be 2.58. This method is reasonable when $\bar{x}$ is associated with a normal distribution with mean μ and standard deviation $SE_{\bar{x}}$. Thus, the formula for a 99% confidence interval for the mean when the Central Limit Theorem applies is

$$\bar{x} \pm 2.58 * SE_{\bar{x}} \tag{4.5}$$

The normal approximation is crucial to the precision of the interval in Equation (4.5). Section 4.4 provides a more detailed discussion about when the normal model can safely be applied. We will find that there are a number of important considerations when evaluating whether we can apply the normal model.

---

[14]This is equivalent to asking how often the Z score will be larger than -2.58 but less than 2.58. (Draw the picture!) To determine this probability, we look up -2.58 and 2.58 in the normal probability table (0.0049 and 0.9951). Thus, there is a $0.9951 - 0.0049 \approx 0.99$ probability that the unobserved random variable X will be within 2.58 standard deviations of μ.

Conditions for $\bar{x}$ being nearly normal and SE being accurate
Important conditions to help ensure the sample mean is nearly normal and the estimate of SE sufficiently accurate:

- The sample observations are independent.
- The sample size is large: $n \geq 50$ is a good rule of thumb.
- The distribution of sample observations is not strongly skewed.

Additionally, the larger the sample size, the more lenient we can be with the sample's skew.

Verifying independence is often the most difficult of the conditions to check, and the way to check for independence varies from one situation to another. However, we can provide simple rules for the most common scenarios.

TIP: How to verify sample observations are independent
Observations in a simple random sample consisting of less than 10% of the population are independent.

TIP: Independence for random processes and experiments
If a sample is from a random process or experiment, it is important to (somehow) verify the observations from the process or subjects in the experiment are nearly independent and maintain their independence throughout.

⊙ **Exercise 4.12** Create a 99% confidence interval for the average age of all runners in the 2009 Cherry Blossom Run. The point estimate is $\bar{y} = 35.22$ and the standard error is $SE_{\bar{y}} = 1.09$. The interval is in the footnote[15].

Confidence interval for any confidence level
If the point estimate follows the normal model with standard error SE, then a confidence interval for the the population parameter is

$$\text{point estimate} \pm z^* SE$$

where z^* corresponds to the confidence level selected.

[15]We verified the conditions earlier (though do it again for practice), so the normal approximation and estimate of SE should be reasonable. We can apply the 99% confidence interval formula: $\bar{y} \pm 2.58 * SE_{\bar{y}} \rightarrow (32.4, 38.0)$. We are 99% confident that the average age of all runners is between 32.4 and 38.0 years.

Figure 4.9 provides a picture of how to identify z^* based on a confidence level. We select z^* so that the area between $-z^*$ and z^* in the normal model corresponds to the confidence level.

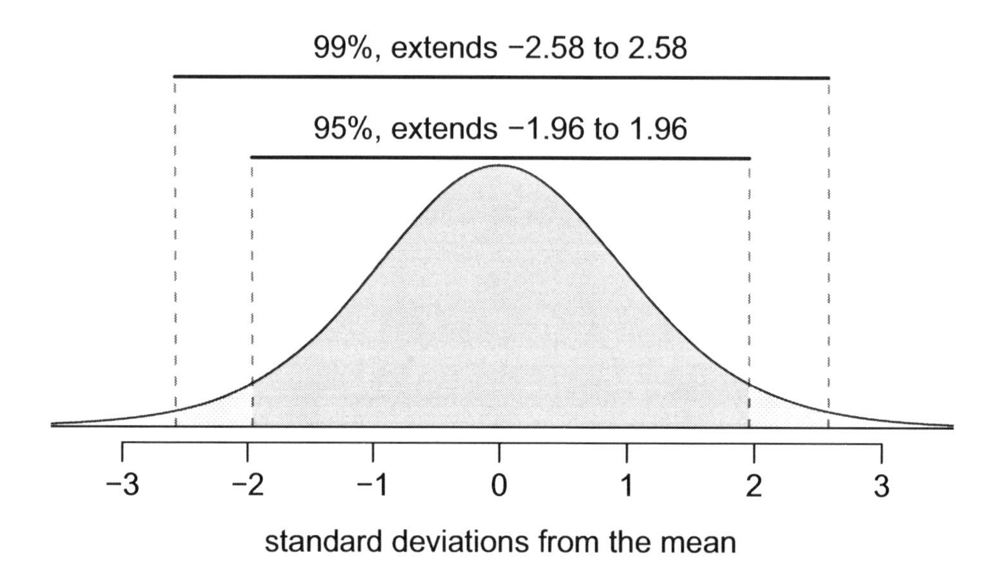

Figure 4.9: The area between $-z^*$ and z^* increases as $|z^*|$ becomes larger. If the confidence level is 99%, we choose z^* such that 99% of the normal curve is between $-z^*$ and z^*, which corresponds to 0.5% in the lower tail and 0.5% in the upper tail; for a 99% confidence level, $z^* = 2.58$.

⊙ **Exercise 4.13** Use the data in Exercise 4.12 to create a 90% confidence interval for the average age of all runners in the 2009 Cherry Blossom Run. Answer in the footnote[16].

4.2.5 Interpreting confidence intervals

A careful eye might have observed the somewhat awkward language used to describe confidence intervals. Correct interpretations:

We are XX% confident that the population parameter is between...

Incorrect language might try to describe the confidence interval as capturing the population parameter with a certain probability. This is one of the most common errors: while it might be useful to *think* of it as a probability, the confidence level only quantifies how plausible it is that the parameter is in the interval.

[16] We need to find z^* such that 90% of data falls between $-z^*$ and z^* in the normal model. We can look up $-z^*$ in the normal probability table by looking for a lower tail of 5% (the other 5% is in the upper tail), thus $z^* = 1.65$. The 90% confidence interval can then be computed as (33.4, 37.0). (We had already verified conditions for normality and the standard error.) That is, we are 90% confident the average age is larger than 33.4 years but less than 37.0 years.

Another especially important consideration of confidence intervals is that they *only try to capture the population parameter.* Our intervals say nothing about the confidence of capturing individual observations, a proportion of the observations, or about capturing point estimates. Confidence intervals only attempt to capture population parameters.

4.3 Hypothesis testing

Is the typical US runner getting faster or slower over time? We consider this question in the context of the Cherry Blossom Run, comparing runners in 2006 and 2009. Technological advances in shoes, training, and diet might suggest runners would be faster in 2009. An opposing viewpoint might say that with the average body mass index on the rise, people tend to run slower. In fact, all of these components might be influencing run time.

In addition to considering run times in this section, we consider a topic near and dear to most college students: sleep. A recent study found that college students average about 7 hours of sleep per night[17]. However, researchers at a rural college believe that their school is different and that their students sleep longer than seven hours on average. We investigate this matter in Section 4.3.4.

4.3.1 Hypothesis testing framework

In 2006, the average time for the Cherry Blossom 10 Mile Run was 93.29 minutes (93 minutes and about 17 seconds). We want to determine if the `run10Samp` data set provides convincing evidence that the participants in 2009 were faster, slower, or the same speed as those runners in 2006[18]. We simplify these three options into two competing **hypotheses**:

H_0: The average 10 mile run time was the same for 2006 and 2009.

H_A: The average 10 mile run time for 2009 was *different* than that of 2006.

We call H_0 the null hypothesis and H_A the alternative hypothesis.

Null and alternative hypotheses
The **alternative hypothesis** represents a claim under consideration and is denoted by H_A. The **null hypothesis** often represents a skeptical perspective relative to the alternative, and it is denoted by H_0.

In hypothesis testing, the null hypothesis often represents a skeptical position or a perspective of no difference. Why should we think run times have actually

[17] http://media.www.theloquitur.com/media/storage/paper226/news/2000/10/19/News/Poll-Shows.College.Students.Get.Least.Amount.Of.Sleep-5895.shtml

[18] While we could answer this question by examining the `run10` population data, we only consider the sample data in `run10Samp`.

changed from 2006 to 2009? The alternative hypothesis often represents a new perspective, such as the claim that there has been a change.

> **TIP: Hypothesis testing framework**
> The hypothesis testing framework is built for a skeptic to consider a new claim. The skeptic will not reject the null hypothesis (H_0), unless the evidence in favor of the alternative hypothesis (H_A) is so convincing that she rejects H_0 in favor of H_A.

The hypothesis testing framework is a very general tool, and we often use it without a second thought. If a person makes a somewhat unbelievable claim, we are initially skeptical. However, if there is sufficient evidence that supports the claim, we set aside our skepticism and reject the null hypothesis in favor of the alternative. The hallmarks of hypothesis testing are also found in the US court system.

⊙ **Exercise 4.14** A US court considers two possible claims about a defendant: she is either innocent or guilty. If we set these claims up in a hypothesis framework, which would be the null hypothesis and which the alternative? Answer in the footnote[19].

Jurors examine the evidence to see whether it convincingly shows a defendant is guilty. Even if the jurors leave unconvinced of guilt beyond a reasonable doubt, this does not mean they believe the defendant is innocent. So it also is with hypothesis testing: *even if we fail to reject the null hypothesis for the alternative, we typically do not accept the null hypothesis as true.* For this reason, failing to find convincing evidence for the alternative hypothesis is not equivalent to accepting the null hypothesis.

In our example with the Cherry Blossom Run, the null hypothesis represents no difference in the average time from 2006 to 2009. The alternative hypothesis represents something new or more interesting: there was a difference, either an increase or a decrease. These hypotheses can be described in mathematical notation using μ_{09} as the average run time for 2009:

H_0: $\mu_{09} = 93.29$

H_A: $\mu_{09} \neq 93.29$

where 93.29 minutes (93 minutes and about 17 seconds) is the average 10 mile time for all runners in 2006 Cherry Blossom Run. In this mathematical notation, our hypotheses can now be evaluated in the context of our statistical tools. We call 93.29 the **null value** since it represents the value of the parameter if the null hypothesis is true.

[19]The jury considers whether the evidence is so convincing that innocence is too implausible; in such a case, the jury rejects innocence (the null hypothesis) and concludes the defendant is guilty (alternative hypothesis).

We will use the `run10Samp` data set to evaluate the hypothesis test. If these data provide convincing evidence that the null hypothesis is false and the alternative is true, then we will reject H_0 in favor of H_A. If we are left unconvinced, then we do not reject the null hypothesis.

4.3.2 Testing hypotheses with confidence intervals

We can start our evaluation of the hypothesis setup comparing 2006 and 2009 run times by using a point estimate: $\bar{x}_{09} = 93.65$ minutes. This estimate suggests the average time is actually longer than the 2006 time, 93.29 minutes. However, to evaluate whether this provides convincing evidence that there has been a change, we must consider the uncertainty associated with $\bar{x}_{09}$.

We learned in Section 4.1 that there is fluctuation from one sample to another, and it is very unlikely that sample mean will be exactly equal to our parameter; we should not expect $\bar{x}_{09}$ to exactly equal μ_{09}. With only knowledge of the `run10Samp`, it might still be possible that the population average in 2009 has remained unchanged from 2006. The difference between $\bar{x}_{09}$ and 93.29 could be due to *sampling variation*, i.e. the variability associated with the point estimate.

In Section 4.2, confidence intervals were introduced as a way to find a range of plausible values for the population mean. Based on `run10Samp`, a 95% confidence interval for the 2009 population mean, μ_{09}, was found:

$$(90.51, 96.79)$$

Because the 2006 mean (93.29) falls in the range of plausible values, we cannot say the null hypothesis is implausible. That is, we fail to reject the null hypothesis, H_0.

> **TIP: Double negatives can sometimes be used in statistics**
> In many statistical explanations, we use double negatives. For instance, we might say that the null hypothesis is *not implausible*. Double negatives are used to communicate that while we are not rejecting a position, we are also not saying it is correct.

● **Example 4.2** We can also consider whether there is evidence that the average age of runners has changed from 2006 to 2009 in the Cherry Blossom Run. In 2006, the average age was 36.13 years, and in the 2009 `run10Samp` data set, the average was 35.22 years with a standard deviation of 10.93 years over 100 runners.

First, we must set up the hypotheses:

H_0: The average age of runners in 2006 and 2009 is the same, $\mu_{age} = 36.13$.

H_A: The average age of runners in 2006 and 2009 is *not* the same, $\mu_{age} \neq 36.13$.

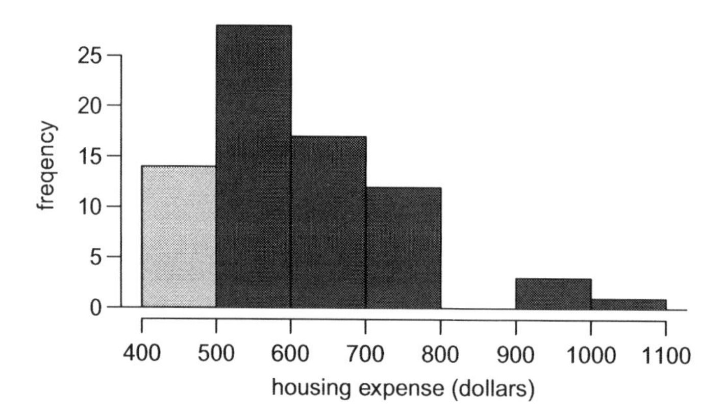

Figure 4.10: Sample distribution of student housing expense.

We have previously verified assumptions on this data set, so we can safely apply the normal model to $\bar{y}$ and compute SE. We can construct a 95% confidence interval for μ_{age} based on the sample to determine if there is convincing evidence to reject H_0:

$$\bar{y} \pm 1.96 * \frac{s_{09}}{\sqrt{100}} \quad \rightarrow \quad 35.22 \pm 1.96 * 1.09 \quad \rightarrow \quad (33.08, 37.36)$$

This confidence interval contains the *null parameter*, 36.13. Because 36.13 is not implausible, we cannot reject the null hypothesis. The sample mean of 35.22 years could easily have occurred even if the actual mean was 36.13, so we have not found convincing evidence that the average age is different than 36.13 years.

⊙ **Exercise 4.15** Colleges frequently give estimates of student expenses such as housing. A community college estimated that the average student housing expense was \$650 per month. Set up hypotheses to evaluate whether the true average housing expense is \$650 per month. [20]

⊙ **Exercise 4.16** The community college decides to collect data to evaluate the \$650 per month claim. They take a random sample of 75 students at their school and obtain the data represented in Figure 4.10. Can we apply the normal model to the sample mean? [21]

● **Example 4.3** The sample mean for student housing is \$611.63 and the sample standard deviation is \$132.85. Construct a 95% confidence interval for the population mean and evaluate the hypotheses of Exercise 4.15.

[20] H_0: The average cost is \$650 per month, $\mu = \$650$.
H_A: The average cost is different than \$650 per month, $\mu \neq \$650$.

[21] Applying the normal model requires that certain conditions are met. Because the data is a simple random sample and the sample (presumably) represents no more than 10% of all students at the college, the observations are independent. The sample size is also sufficiently large ($n = 75$) and the data are skewed but not strongly skewed. Thus, the normal model may be applied to the sample mean.

		Test conclusion	
		do not reject H_0	reject H_0 in favor of H_A
Truth	H_0 true	okay	Type 1 Error
	H_A true	Type 2 Error	okay

Table 4.11: Four different scenarios for hypothesis tests.

The standard error associated with the mean may be estimated using the sample standard deviation divided by the square root of the sample size. Recall that $n = 75$ students were sampled.

$$SE = \frac{s}{\sqrt{n}} = \frac{132.85}{\sqrt{75}} = 15.34$$

You showed in Exercise 4.16 that the normal model may be applied to the sample mean. This ensures a 95% confidence interval may be accurately constructed:

$$\bar{x} \pm z^* SE \quad \rightarrow \quad 611.63 \pm 1.96 * (15.34) \quad \rightarrow \quad (581.56, 641.70)$$

Because the null value \$650 is not in the confidence interval, we conclude that \$650 is a statistically implausible mean in light of the data and we reject the null hypothesis. We have statistically significant evidence to conclude that the actual housing expense averages less than \$650 per month.

4.3.3 Decision errors

Are hypothesis tests flawless? Absolutely not. Just think of the courts system: innocent people are sometimes wrongly convicted and the guilty sometimes walk free. Similarly, we can make a wrong decision in statistical hypothesis tests as well. However, the difference is that we have the tools necessary to quantify how often we make errors in statistics.

There are two competing hypotheses: the null and the alternative. In a hypothesis test, we make some commitment about which one might be true, but we might choose incorrectly. There are four possible scenarios in a hypothesis test. They are summarized in Table 4.11.

A **Type 1 Error** is rejecting the null hypothesis when H_0 is actually true. A **Type 2 Error** is failing to reject the null hypothesis when the alternative is actually true.

⊙ **Exercise 4.17** In a US court, the defendant is either innocent (H_0) or guilty (H_A). What does a Type 1 Error represent in this context? What does a Type 2 Error represent? Table 4.11 may be useful. [22]

[22] If the court makes a Type 1 Error, this means the defendant is innocent (H_0 true) but wrongly convicted, i.e. the court wrongly rejects H_0. A Type 2 Error means the court failed to reject H_0 (i.e. fail to convict the person) when she was in fact guilty (H_A true).

⊙ **Exercise 4.18** How could we reduce the Type 1 Error rate in US courts? What influence would this have on the Type 2 Error rate? [23]

⊙ **Exercise 4.19** How could we reduce the Type 2 Error rate in US courts? What influence would this have on the Type 1 Error rate?

Exercises 4.17-4.19 provide an important lesson: if we reduce how often we make one type of error, we generally make more of the other type.

Hypothesis testing is built around rejecting or failing to reject the null hypothesis. That is, we do not reject H_0 without convincing evidence. But what precisely do we mean by *convincing evidence*? As a general rule of thumb, for those cases where the null hypothesis is actually true, we do not want to incorrectly reject H_0 more than 5% of those times. This corresponds to a **significance level** of 0.05. We often write the significance level using α (the Greek letter *alpha*): $\alpha = 0.05$. Selection considerations for an appropriate significance level are discussed in Section 4.3.6.

If we use a 95% confidence interval to test a hypothesis, then we will make an error whenever the point estimate is at least 1.96 standard errors away. This happens about 5% of the time (2.5% in each tail). Similarly, if we used a 99% confidence interval to evaluate our hypothesis, the test level would be $\alpha = 0.01$.

However, a confidence interval is, in one sense, simplistic in the world of hypothesis tests. Consider the following two scenarios:

- The parameter value under the null hypothesis is in the 95% confidence interval but just barely, so we would not reject H_0. However, we might like to somehow say, quantitatively, that it was a close decision.

- The null value is very far outside of the interval, so we reject H_0. However, we want to communicate that, not only did we reject the null hypothesis, but it wasn't even close. Such a case is depicted in Figure 4.12.

In Section 4.3.4, we introduce a tool called the *p-value* that will be helpful in these cases. The p-value method also extends to hypothesis tests where confidence intervals cannot be easily constructed or applied.

4.3.4 Formal testing using p-values

The p-value is a way to quantify how the strength of the evidence against the null hypothesis and in favor of the alternative. Formally we define the *p-value* as a conditional probability.

[23] To lower the Type 1 Error rate, we might raise our standard for conviction from "beyond a reasonable doubt" to "beyond a conceivable doubt" so fewer people would be wrongly convicted. However, this would also make it more difficult to convict the people who are actually guilty, i.e. we will make more Type 2 Errors.

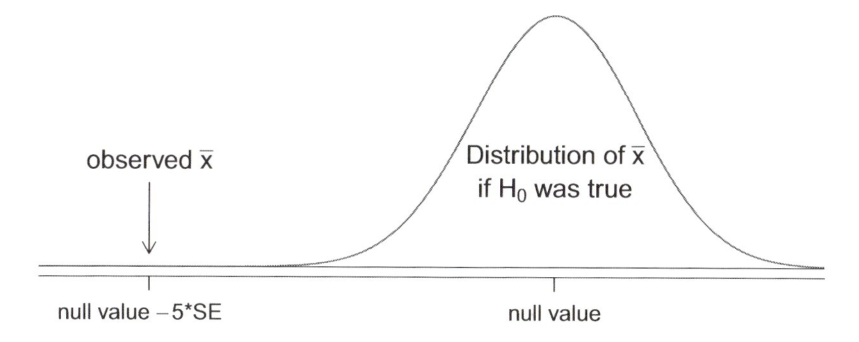

Figure 4.12: It would be helpful to quantify the strength of the evidence against the null hypothesis. In this case, the evidence is extremely strong.

> **p-value**
> The **p-value** is the probability of observing data at least as favorable to the alternative hypothesis as our current data set, if the null hypothesis was true. We typically use a summary statistic of the data, in this chapter the sample mean, to help us compute the p-value and evaluate the hypotheses.

A poll by the National Sleep Foundation found that college students average about 7 hours of sleep per night. Researchers at a rural school believe that students at their school sleep longer than seven hours on average, and they would like to demonstrate this using a sample of students. There are three possible cases: the students at the rural school average more than 7 hours of sleep, they average less than 7 hours, or they average exactly 7 hours.

⊙ **Exercise 4.20** What would be a skeptical position in regards to the researchers' claims? [24]

We can set up the null hypothesis for this test as a skeptical perspective of the researchers' claim: the students at this school average 7 or fewer hours of sleep per night. The alternative hypothesis takes a new form reflecting the purpose of the study: the students average more than 7 hours of sleep. We can write these hypotheses in statistical language:

$H_0: \mu \leq 7.$

$H_A: \mu > 7.$

While we might think about the null hypothesis as representing a range of values in this case, it is common to simply write it as an equality:

[24] A skeptic would have no reason to believe that the researchers are correct: maybe the students at this school average 7 or fewer hours of sleep.

$H_0 : \mu = 7.$

$H_A : \mu > 7.$

This new hypothesis setup with $\mu > 7$ as the alternative is an example of a **one-sided** hypothesis test. In this investigation we are not concerned with seeing whether the mean is less than 7 hours[25]. Our earlier hypothesis setups are termed **two-sided** since they looked for any clear difference, greater than or less than.

Always use a two-sided test unless it was made clear prior to data collection that the test should be one-sided. Switching a two-sided test to a one-sided test after observing the data is dangerous because it can inflate the Type 1 Error rate.

TIP: One-sided and two-sided tests
If the researchers are only interested in showing *either* an increase *or* a decrease, then set up a one-sided test. If the researchers would be interested in any difference from the null value – an increase or decrease – then the test should be two-sided.

TIP: Always write the null hypothesis as an equality
We will find it most useful if we always list the null hypothesis as an equality (e.g. $\mu = 7$) while the alternative always uses an inequality (e.g. $\mu \neq 7$, $\mu > 7$, or $\mu < 7$).

The researchers conducted a simple random sample of $n = 110$ students on campus. They found that these students averaged 7.42 hours of sleep and the standard deviation of the amount of sleep for the students was 1.75 hours. A histogram of the sample is shown in Figure 4.13.

Before we can use a normal model for the sample mean or compute the standard error of the sample mean, we must verify conditions. (1) Because the sample is simple random from less than 10% of the student body, the observations are independent. (2) The sample size in the sleep study is sufficiently large (greater than 50). (3) The data is not too strongly skewed in Figure 4.13. With these conditions verified, we can rest assured that the normal model can be safely applied to $\bar{x}$ and we can dependably compute the standard error.

⊙ **Exercise 4.21** What is the standard deviation associated with $\bar{x}$? That is, estimate the standard error of $\bar{x}$.[26]

[25] This is entirely based on the interests of the researchers. Had they been only interested in the opposite case – showing that their students were actually averaging fewer than seven hours of sleep but not interested in showing more than 7 hours – then our setup would have set the alternative as $\mu < 7$.

[26] The standard error can be estimated from the sample standard deviation and the sample size: $SE_{\bar{x}} = \frac{s_x}{\sqrt{n}} = \frac{1.75}{\sqrt{110}} = 0.17$.

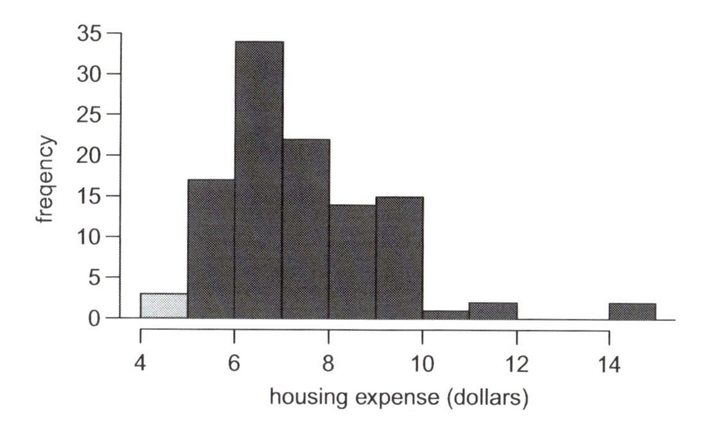

Figure 4.13: Distribution of a night of sleep for 110 college students.

We will evaluate the hypotheses using a significance level of $\alpha = 0.05$. We want to consider the data under the scenario that the null hypothesis is true. In this case, the sample mean is from a distribution with mean 7 and standard deviation of about 0.17 that is nearly normal such as the one shown in Figure 4.14.

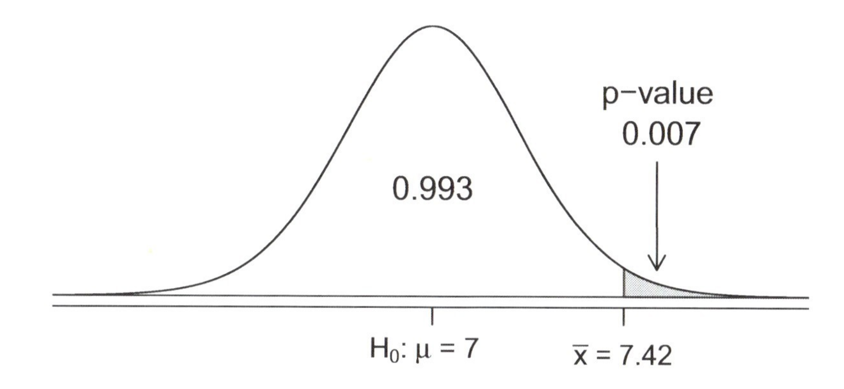

Figure 4.14: If the null hypothesis was true, then the sample mean $\bar{x}$ came from this nearly normal distribution. The right tail describes the probability of observing such a large sample mean if the null hypothesis was true.

The shaded tail in Figure 4.14 represents the chance of observing such an unusually large mean, conditioned on the null hypothesis being true. That is, the shaded tail represents the p-value. We shade all means larger than our sample mean, $\bar{x} = 7.42$, because they are more favorable to the alternative hypothesis.

We can compute the p-value by finding this tail area of this normal distribution, which we learned to do in Section 3.1. We first compute the Z score of the sample mean, $\bar{x} = 7.42$:

$$Z = \frac{\bar{x} - \mu_0}{SE_{\bar{x}}} = \frac{7.42 - 7}{0.17} = 2.47$$

Using the normal probability table, the lower unshaded area is found to be 0.993. Thus the shaded area is $1 - 0.993 = 0.007$. *If the null hypothesis was true, the probability of observing such a large mean is only 0.007.* That is, if the null hypothesis was true, we would not often see such a large mean.

We evaluate the hypotheses by comparing the p-value to the significance level. Because the p-value is less than the significance level (p-value $= 0.007 < 0.050 = \alpha$), we reject the null hypothesis. What we observed is so unusual with respect to the null hypothesis that it casts serious doubt on H_0 and provides strong evidence favoring H_A.

p-value as tool in hypothesis testing
The p-value quantifies how strongly the data favors H_A over H_0. A small p-value (usually < 0.05) corresponds to convincing evidence to reject H_0.

TIP: It is useful to first draw a picture to find the p-value
It is useful to draw a picture of the distribution of $\bar{x}$ if H_0 was true (e.g. μ equals the null value), and shade the region (or regions) of sample means that are even more favorable to the alternative hypothesis. These shaded regions represent the p-value.

The ideas below review the process of evaluating hypothesis tests with p-values:

- The null hypothesis represents a skeptic's position. We reject this position only if the evidence convincingly favors H_A.
- A small p-value means that if the null hypothesis was true, there is a low probability of seeing a point estimate at least as extreme as the one we saw. We interpret this as strong evidence in favor of the the alternative.
- We reject the null hypothesis if the p-value is smaller than the significance level, α, which is usually 0.05. Otherwise, we fail to reject H_0.
- We should always state the conclusion of the hypothesis test in plain language so non-statisticians can also understand the results.

The p-value is constructed in such a way that we can directly compare it to the significance level, α, to determine whether we should or should not reject H_0. This method ensures that the Type 1 Error does not exceed the significance level standard.

⊙ **Exercise 4.22** If the null hypothesis is true, how often should the test p-value be less than 0.05? Answer in the footnote[27].

[27] About 5% of the time. If the null hypothesis is true, then the data only has a 5% chance of being in the 5% of data most favorable to H_A.

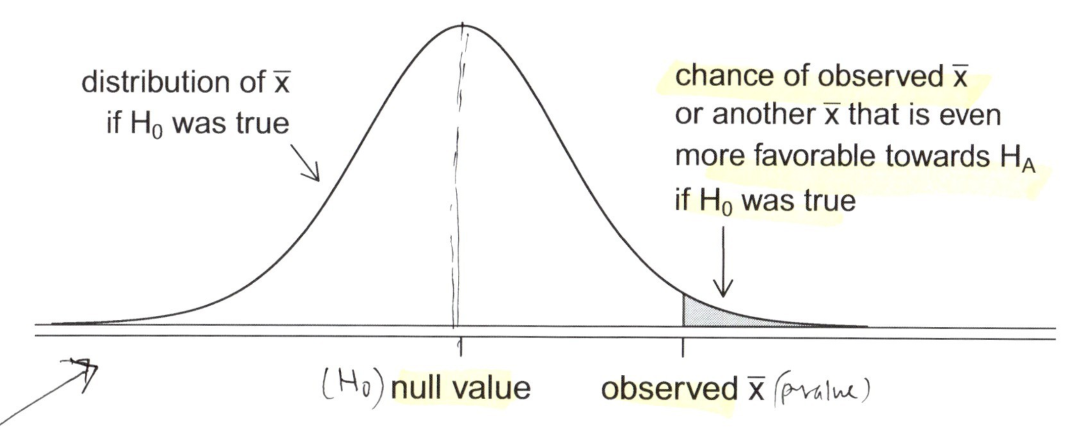

Figure 4.15: To identify the p-value, the distribution of the sample mean is considered if the null hypothesis was true. Then the p-value is defined and computed as the probability of the observed $\bar{x}$ or an $\bar{x}$ even more favorable to H_A under this distribution.

⊙ **Exercise 4.23** Suppose we had used a significance level of 0.01 in the sleep study. Would the evidence have been convincing enough to reject the null hypothesis? (The p-value was 0.007.) What if the significance level was $\alpha = 0.001$? Answers in the footnote[28].

⊙ **Exercise 4.24** We would like to evaluate whether Ebay buyers on average pay less than the price on Amazon. During early October 2009, Amazon sold a game called *Mario Kart* for the Nintendo Wii for \$46.99. Set up an appropriate (one-sided!) hypothesis test to check the claim that Ebay buyers pay less. Answer in the footnote[29].

⊙ **Exercise 4.25** During early October, 2009, 52 Ebay auctions were recorded for *Wii Mario Kart*. We would like to apply the normal model to the sample mean, however, we must verify three conditions to do so: (1) independence, (2) at least 50 observations, and (3) the data are not extremely skewed. Assume the last condition is satisfied. Do you think the first two conditions would be reasonably met? [30]

---

[28]We reject the null hypothesis whenever *p-value* $< \alpha$. Thus, we would still reject the null hypothesis if $\alpha = 0.01$. However, we would not have sufficient evidence to reject the null hypothesis had the significance level been $\alpha = 0.001$.

[29]The skeptic would say the average is the same (or more) on Ebay, and we are interested in showing the average price is lower.

H_0: The average auction price on Ebay is equal to (or more than) the price on Amazon. We write only the equality in the statistical notation: $\mu_{ebay} = 46.99$.

H_A: The average price on Ebay is less than the price on Amazon, $\mu_{ebay} < 46.99$.

[30]We should first verify assumptions to ensure we can apply the normal model. (1) The independence condition is unclear. *We will make the assumption that the observations are independence*

● **Example 4.4** The average sale price of the 52 Ebay auctions for *Wii Mario Kart* was \$44.17 with a standard deviation of \$4.15. Does this provide sufficient evidence to reject the null hypothesis in Exercise 4.24? Use a significance level of $\alpha = 0.01$ and assume the conditions for applying the normal model are satisfied.

The hypotheses were set up and the conditions were checked in Exercises 4.24 and 4.25. The next step is to find the standard error of the sample mean and produce a sketch to help find the p-value.

$$SE_{\bar{x}} = s/\sqrt{n} = 4.15/\sqrt{52} = 0.5755$$

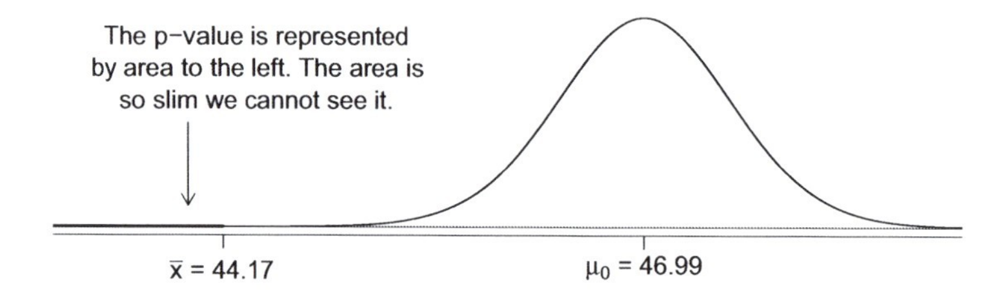

Because the alternative hypothesis says we are looking for a smaller mean, we shade the lower tail. We find this shaded area by using the Z score and normal probability table: $Z = \frac{44.17-46.99}{0.5755} = -4.90$, which has area less than 0.0002. The area is so small we cannot really see it on the picture. This lower tail area corresponds to the p-value.

Because the p-value is so small – specifically, smaller than $\alpha = 0.01$ – this provides convincing evidence to reject the null hypothesis in favor of the alternative. The data provides statistically significant evidence that the average price on Ebay is lower than Amazon's asking price.

4.3.5 Two-sided hypothesis testing with p-values

We now consider how to compute a p-value for a two-sided test. In one-sided tests, we shade the single tail in the direction of the alternative hypothesis. For example, when the alternative had the form $\mu > 7$, then the p-value was represented by the upper tail (Figure 4.15). When the alternative was $\mu < 46.99$, the p-value was the lower tail (Exercise 4.25). In a two-sided test, *we shade two tails* since any observation that is more favorable to H_A than what we observed contributes to the p-value.

⊙ **Exercise 4.26** Earlier we talked about a research group that examined whether the students at their school slept longer than 7 hours each night. Let's consider a second group of researchers that wants to evaluate whether the students at a second college match the norm. They don't have any expectations and would simply like to

holds. (2) The sample size is sufficiently large: $n = 52 \geq 50$. (3) While not shown, the distribution of the sample is nearly symmetric. Under the assumption of independence, we use the normal model.

see whether the data suggests the students match the 7 hour average. Write out the null and alternative hypotheses for this investigation. [31]

● **Example 4.5** The second college randomly samples 72 students and finds a mean of $\bar{x} = 6.83$ hours and a standard deviation of $s = 1.8$ hours. Does this provide convincing evidence against H_0 in Exercise 4.26? Use a significance level of $\alpha = 0.05$.

First, we must verify assumptions. (1) A simple random sample of less than 10% of the student body means the observations are independent. (2) The sample size is greater than 50. (3) Based on the earlier distribution and what we already know about college student sleep habits, the distribution is probably not too strongly skewed.

Next we can compute the standard error ($SE_{\bar{x}} = \frac{s}{\sqrt{n}} = 0.21$) of the estimate and creating a picture to represent the p-value, shown in Figure 4.16. Both tails are shaded since anything larger than 7.17 provides at least as much evidence against H_0 and in favor of H_A as 6.83. That is, an estimate of 7.17 or more provides at least as strong of evidence against the null hypothesis and in favor of the alternative as the observed estimate, $\bar{x} = 6.83$.

We can find the tail areas by first finding the lower tail corresponding to $\bar{x}$:

$$Z = \frac{6.83 - 7.00}{0.21} = -0.81 \quad \overset{table}{\rightarrow} \quad \text{left tail} = 0.2090$$

Because the normal model is symmetric, the right tail will have the same area as the left tail. The p-value is found as the sum of the two shaded tails:

$$\text{p-value} = \text{left tail} + \text{right tail} = 2 * (\text{left tail}) = 0.4180$$

This p-value is relatively large (larger than $\alpha = 0.05$) so we should not reject H_0. That is, if H_0 was true it would not be very unusual to see a sample mean this far from 7 hours simply due to sampling variation. Thus, we do not have convincing evidence that the mean is different from 7 hours.

● **Example 4.6** It is never okay to change two-sided tests to one-sided tests after observing the data. In this example we explore the consequences of ignoring this advice. Use $\alpha = 0.05$ we show that freely switching from two-sided tests to one-sided tests will cause us to make twice as many Type 1 Errors than we intend.

Suppose we observe $\bar{x}$ above the null value, which we label as μ_0 (e.g. μ_0 would represent 7 if $H_0 : \mu = 7$). Then if we can flip to a one-sided test, we would use $H_A : \mu > \mu_0$. Now if we would obtain any observation with a Z score greater than 1.65, we would reject H_0. If the null hypothesis is true, we incorrectly reject the null hypothesis about 5% of the time when the sample mean is above the null value, as shown in Figure 4.17.

[31]Because the researchers are interested in any difference, they should use a two-sided setup: $H_0 : \mu = 7$, $H_A : \mu \neq 7$.

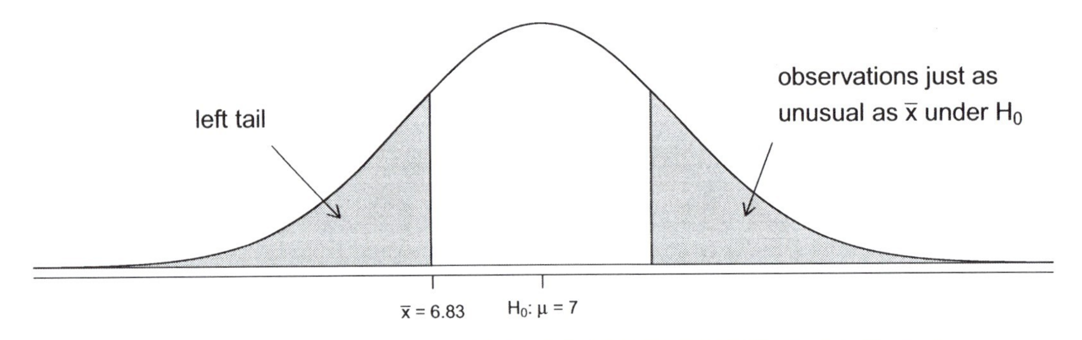

Figure 4.16: H_A is two-sided, so *both* tails must be counted for the p-value.

Suppose we observe $\bar{x}$ below the null value, μ_0. Then if we can flip to a one-sided test, we would use $H_A : \mu < \mu_0$. If $\bar{x}$ had a Z score smaller than -1.65, we would reject H_0. If the null hypothesis was true, then we would observe such an case about 5% of the time.

By examining these two scenarios, we can determine that we will make a Type 1 Error 5% + 5% = 10% of the time if we are allowed to swap in the "best" one-sided test for the data. This is twice the error rate we prescribed with our significance level: $\alpha = 0.05$ (!).

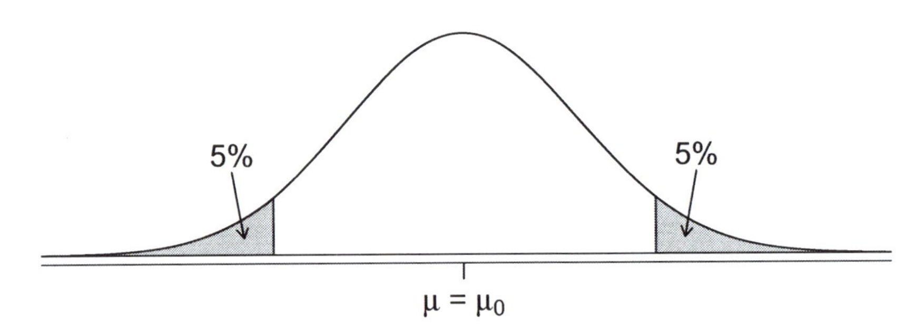

Figure 4.17: The shaded regions represent areas where we would reject H_0 under the bad practices considered in Example 4.6 when $\alpha = 0.05$.

Caution: One-sided tests allowed only *before* seeing data
After observing data, it is tempting to turn a two-sided test into a one-sided test. Avoid this temptation. Hypotheses must be set up *before* observing the data. If they are not, the test must be two-sided.

4.3.6 Choosing a significance level

Choosing a significance level for a test is especially important in many contexts. The traditional level is 0.05. However, because we would like to apply our techniques to a variety of scenarios, it is helpful to choose significance levels appropriate to the context. We may select a level that is smaller or larger than 0.05 depending on the consequences of any conclusions reached from the test.

If making a Type 1 Error is dangerous or especially costly, we should choose a small significance level (e.g. 0.01). Under this scenario we want to be very cautious about rejecting the null hypothesis, so we demand very strong evidence to reject H_0.

If a Type 2 Error is relatively more dangerous or much more costly than a Type 1 Error, then we should choose a higher significance level (e.g. 0.10). Here we want to be cautious about failing to reject H_0 when the null is actually false.

Significance levels should reflect consequences of errors
The significance level selected for a test should reflect the consequences associated with Type 1 and Type 2 Errors.

● **Example 4.7** A car manufacturer is considering a higher quality but more expensive supplier for window parts in its vehicles. They sample a number of parts from their current supplier and also parts from the new supplier. They decide that if the durability is more than 10% longer in the higher quality parts, it makes financial sense to switch to this more expensive supplier. Is there good reason to modify the significance level in such a hypothesis test?

The null hypothesis is that the more expensive goods last no more than 10% longer while the alternative is that they do last more than 10% longer. This decision is unlikely to have an enormous impact on the success of the car or company. A significance level of 0.05 seem reasonable since neither a Type 1 or Type 2 error should be dangerous or (relatively) much more expensive.

● **Example 4.8** The same car manufacturer is considering a slightly more expensive supplier for parts related to safety, not windows. If the durability of these safety components is shown to be better than the current supplier, they will switch manufacturers. Is there good reason to modify the significance level in such an evaluation?

The null hypothesis would be that the suppliers' parts are equally reliable. Because safety is involved, the car company should be eager to switch to the slightly more expensive manufacturer (reject H_0) if the evidence is only moderately strong. A slightly larger significance level, such as $\alpha = 0.10$, might be appropriate.

⊙ **Exercise 4.27** A part inside of a machine is very expensive to replace. However, the machine usually functions properly even if this part is broken so we decide to replace the part only if we are extremely certain it is broken based on a series of measurements. Identify appropriate hypotheses (in words) and suggest an appropriate significance level. [32]

[32]Here the null hypothesis is that the part is not broken. If we don't have sufficient evidence

4.4 Examining the Central Limit Theorem

The normal model for the sample mean tends to be very good when the sample consists of at least 50 independent observations and the original data are not strongly skewed. The Central Limit Theorem provides the theory that allows us to make this assumption.

Central Limit Theorem, informal definition
The distribution of $\bar{x}$ is approximately normal. The approximation can be poor if the sample size is small but gets better with larger sample sizes.

The Central Limit Theorem states that when the sample size is small, the normal approximation may not be very good. However, as the sample size becomes large, the normal approximation improves. We will investigate two cases to see roughly when the approximation is reasonable.

We consider two data sets: one from a *uniform* distribution, the other from an *exponential*. These distributions were introduced in Sections 3.5 and 3.6 and are shown in the top panels of Figure 4.18. [Authors' note: these sections will be added at a later date to Chapter 3.]

The left panel in the $n = 2$ row represents the sampling distribution of $\bar{x}$ if it is the sample mean of two observations from the uniform distribution shown. The red dashed line represents the closest approximation of the normal distribution. Similarly, the right panel of the $n = 2$ row represents the distribution of $\bar{x}$ if its observations came from an exponential distribution, and so on.

⊙ **Exercise 4.28** Examine the distributions in each row of Figure 4.18. What do you notice about the distributions as the sample size becomes larger? Answer in the footnote[33].

⊙ **Exercise 4.29** Would the normal approximation be good in all cases where the sample size is at least 50? Answer in the footnote[34].

TIP: With larger n, $\bar{x}$ becomes more normal
As the sample size increases, the normal model for $\bar{x}$ becomes more reasonable. We can also relax our condition on skew when the sample size is very large.

---

to reject H_0, we would not replace the part. It sounds like failing to fix the part if it is broken (H_0 false, H_A true) is not very problematic, and replacing the part is expensive. Thus, we should require very strong evidence against H_0 to be convinced we should replace the part. Choose a small significance level, such as 0.01.

[33]When the sample size is small, the normal approximation is generally poor. However, as the sample size grows, the normal model becomes better for each distribution. The normal approximation is pretty good when $n = 15$ and very good when $n = 50$.

[34]Not necessarily. It took the skewed distribution in Figure 4.18 longer for $\bar{x}$ to be nearly normal. If a distribution is even more skewed, then we would need a larger sample to guarantee the distribution of the sample mean is nearly normal.

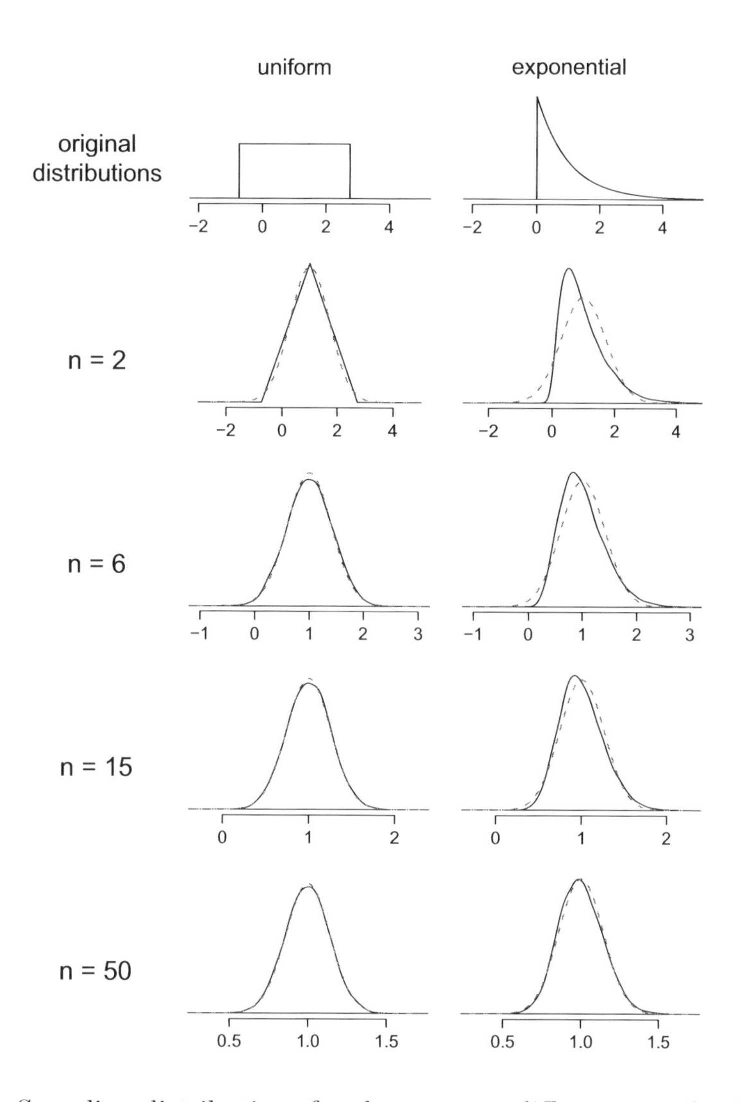

Figure 4.18: Sampling distributions for the mean at different sample sizes and for two data distributionss.

We also discussed in Section 4.1.3 that the sample standard deviation, s, could be used as a substitute of the population standard deviation, σ, when computing standard error. This estimate tends to be reasonable when $n \geq 50$. When $n < 50$, we must use small sample inference tools, which we examine in Chapter 6.

Independence & $n \geq 50$ & no strong skew $\rightarrow \bar{x}$ nearly normal
When we have at least 50 independent observations and the data do not show strong skew, then we can assume $\bar{x}$ follows the normal approximation and our estimate of standard error is sufficiently good.

● **Example 4.9** Figure 4.19 shows a histogram of 50 observations. These represent winnings and losses from 50 consecutive days of a professional poker player. Can the normal approximation be applied to the sample mean, 90.69?

We should consider each of our conditions.

(1) Because the data arrives in a particular sequence, this is called *time series data*. If the player wins on one day, it may influence how he plays the next. To make the assumption of independence we should perform careful checks on such data, which we do not discuss here.

(2) The sample size is 50; this condition is satisfied.

(3) The data set is strongly skewed. In strongly skewed data sets, outliers can play an important role and affect the distribution of the sample mean and the estimate of the standard error.

Since we should be skeptical of the independence of observations and the data is very strongly skewed, we should not apply the normal model for the sample mean.

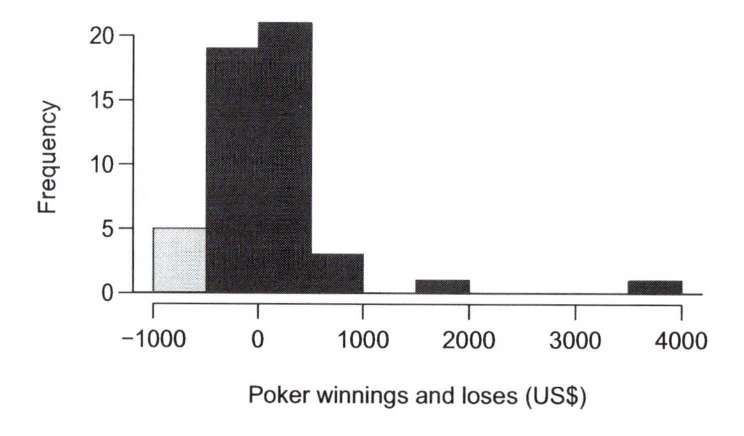

Figure 4.19: Sample distribution of poker winnings.

Caution: Examine the nature of the data when considering independence
Some data sets are collected in such a way that they have a natural underlying structure between observations, most especially consecutive observations. Be especially cautious about making independence assumptions regarding such data sets.

Caution: Watch out for strong skew
If the original data is strongly skewed, the normal approximation may not be appropriate for the sample mean even when the sample size is at least 50.

4.5 A framework for inference

The sample mean is not the only point estimate that is nearly normal. The sampling distribution of sample proportions closely resembles the normal distribution if the sample size is sufficiently large, and so it is also for the difference in two sample means. In this section, we present a general framework for inference that we will use for a wide variety of parameters in Chapter 5. In Chapter 6 and beyond, we look at point estimates that follow other distributions.

For each point estimate in this section, we make two important assumptions: its associated sampling distribution is nearly normal and the estimate is unbiased. A point estimate is **unbiased** if the sampling distribution of the estimate is centered at the parameter it estimates. That is, an unbiased estimate does not naturally over or underestimate the parameter but tends to provide a "good" estimate. The sample mean is an example of an unbiased point estimate.

4.5.1 A general approach to confidence intervals

In Section 4.2, we used the point estimate $\bar{x}$ with a standard error $SE_{\bar{x}}$ to create a 95% confidence interval for the population mean:

$$\bar{x} \pm 1.96 * SE_{\bar{x}} \tag{4.6}$$

We constructed this interval by noting that the sample mean is within 1.96 standard errors of the actual mean about 95% of the time. This same logic generalizes to any unbiased point estimate that is nearly normal.

General 95% confidence interval
If a point estimate is nearly normal and unbiased, then a general 95% confidence interval for the estimated parameter is

$$\text{point estimate} \pm 1.96 * SE \tag{4.7}$$

where SE represents the standard error of the estimate.

Recall that the standard error represents the standard deviation associated with a point estimate. Generally the standard error is estimated from the data and computed using a formula. For example, when the point estimate is a sample mean, we compute the standard error using the sample standard deviation and sample size in the following formula:

$$SE_{\bar{x}} = \frac{s}{\sqrt{n}}$$

Similarly, there is a formula for the standard error of many commonly used point estimates, and we will encounter many of these cases in Chapter 5.

● **Example 4.10** In Exercise 4.3 on page 126, we computed a point estimate for the average difference in run times between men and women: $\bar{x}_{women} - \bar{x}_{men} = 6.82$ minutes. Suppose this point estimate is associated with a nearly normal distribution with standard error $SE = 3.05$ minutes. What is a reasonable 95% confidence interval for the difference in average run times?

The normal approximation is said to be valid and the estimate unbiased, so we apply Equation (4.7):

$$6.82 \pm 1.96 * 3.05 \quad \rightarrow \quad (0.84, 12.80)$$

Thus, we are 95% confident that the men were, on average, between 0.84 and 12.80 minutes faster than women in the 2009 Cherry Blossom Run. That is, the actual average difference is plausibly between 0.84 and 12.80 minutes with 95% confidence.

● **Example 4.11** Does Example 4.10 guarantee that if a husband and wife both ran in the race, the husband would run between 0.84 and 12.80 minutes faster than the wife?

Our confidence interval says absolutely nothing about individual observations. It only makes a statement about a plausible range of values for the *average* difference between all men and women who participated in the run.

The 95% confidence interval for the population mean was generalized for other confidence levels in Section 4.2.4. The constant 1.96 was replaced with some other number z^*, where z^* represented an appropriate Z score for the new confidence level. Similarly, Equation (4.7) can be generalized for a different confidence level.

> **General confidence interval**
> A confidence interval based on an unbiased and nearly normal point estimate is
>
> $$\text{point estimate} \pm z^* SE$$
>
> where SE represents the standard error, and z^* is selected to correspond to the confidence level.

⊙ **Exercise 4.30** What z^* would be appropriate for a 99% confidence level? For help, see Figure 4.9 on page 136.

⊙ **Exercise 4.31** The proportion of men in the `run10Samp` sample is $\hat{p} = 0.48$. This sample meets certain conditions that guarantee $\hat{p}$ will be nearly normal, and the standard error of the estimate is $SE_{\hat{p}} = 0.05$. Create a 90% confidence interval for the proportion of participants in the 2009 Cherry Blossom Run who are men. [35]

4.5.2 Hypothesis testing generalization

Just as we generalized our confidence intervals for any unbiased and nearly normal point estimate, we can generalize our hypothesis testing methods. Here we only consider the p-value approach, introduced in Section 4.3.4, since it is the most commonly used technique and also extends to non-normal cases (Chapter 6).

⊙ **Exercise 4.32** The Food and Drug Administration would like to evaluate whether a drug (sulphinpyrazone) is effective at reducing the death rate in heart attack patients. Patients were randomly split into two groups: a control group that received a placebo and a treatment group that received the new drug. What would be an appropriate null hypothesis? And the alternative? [36]

We can formalize the hypotheses from Exercise 4.32 by letting $p_{control}$ and $p_{treatment}$ represent the proportion of patients who died in the control and treatment groups, respectively. Then the hypotheses can be written as

$$H_0 : p_{control} = p_{treatment} \quad \text{(the drug doesn't work)}$$
$$H_A : p_{control} > p_{treatment} \quad \text{(the drug works)}$$

---

[35] We use $z^* = 1.65$ (see Exercise 4.13 on page 136), and apply the general confidence interval formula:

$$\hat{p} \pm z^* SE_{\hat{p}} \quad \rightarrow \quad 0.48 \pm 1.65 * 0.05 \quad \rightarrow \quad (0.3975, 0.5625)$$

Thus, we are 90% confident that between 40% and 56% of the participants were men.

[36] The skeptic's perspective is that the drug does not work (H_0), while the alternative is that the drug does work (H_A).

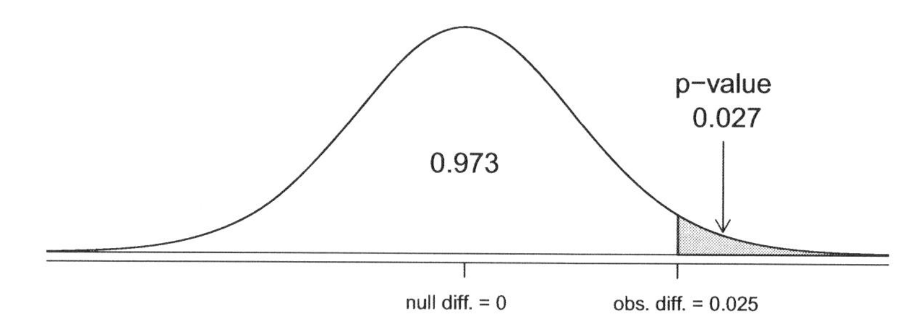

Figure 4.20: The distribution of the sample difference if the null hypothesis was true.

or equivalently,

$$H_0 : p_{control} - p_{treatment} = 0 \quad \text{(the drug doesn't work)}$$
$$H_A : p_{control} - p_{treatment} > 0 \quad \text{(the drug works)}$$

Strong evidence against the null hypothesis and in favor of the alternative would correspond to an observed difference in death rates,

$$\hat{p}_{control} - \hat{p}_{treatment}$$

being larger than we would expect from chance alone. This difference in sample proportions represents a point estimate that is useful in evaluating the hypotheses.

● **Example 4.12** We want to evaluate the hypothesis setup from Exericse 4.32 using data introduced in Section 1.1. In the control group, 60 of the 742 patients died. In the treatment group, 41 of 733 patients died. The sample difference in death rates can be summarized as

$$\hat{p}_{control} - \hat{p}_{treatment} = \frac{60}{742} - \frac{41}{733} = 0.025$$

As we will find out in Chapter 5, we can safely assume this point estimate is nearly normal and is an unbiased estimate of the actual difference in death rates. The standard error of this sample difference is $SE = 0.013$. We would like to evaluate the hypothesis at a 5% significance level: $\alpha = 0.05$.

We would like to identify the p-value to evaluate the hypotheses. If the null hypothesis was true, then the point estimate would have come from a nearly normal distribution, like the one shown in Figure 4.20. The distribution is centered at zero since $p_{control} - p_{treatment} = 0$ under the null hypothesis. Because a large positive difference provides evidence against the null hypothesis and in favor of the the alternative, the upper tail tail has been shaded to represent the p-value. We need not shade the lower tail since this is a one-sided test: an observation in the lower tail does not support the alternative hypothesis.

The p-value can be computed by using a Z score of our point estimate and the normal probability table.

$$Z = \frac{\text{point estimate} - \text{null value}}{SE_{\text{point estimate}}} = \frac{0.025 - 0}{0.013} = 1.92 \tag{4.8}$$

In our Z score formula, we replaced our sample mean with our general point estimate and included a corresponding standard error. Examining Z in the normal probability table, we find that the lower unshaded tail is about 0.973. Thus, the upper shaded tail representing the p-value is

$$\text{p-value} = 1 - 0.973 = 0.027$$

Because the p-value is less than the significance level ($\alpha = 0.05$), we say the null hypothesis is implausible. That is, we reject the null hypothesis in favor of the alternative and conclude that the drug is effective at reducing deaths in heart attack patients.

Provided a point estimate is unbiased and nearly normal, the methods for identifying the p-value and evaluating a hypothesis test change little from Section 4.3.4.

Hypothesis testing using the normal model

1. First set up the hypotheses in plain language, then set it up in mathematical notation.
2. Identify an appropriate point estimate of the parameter in your hypotheses.
3. Verify conditions to ensure the standard error estimate is good and the point estimate is nearly normal and unbiased.
4. Compute its standard error. Draw a picture depicting the distribution of the estimate, if H_0 was true. Shade areas representing the p-value.
5. Using the picture and normal model, compute the *test statistic* (Z score) and identify the p-value to evaluate the hypotheses. Write a conclusion in plain language.

The Z score in Equation (4.8) is called a **test statistic**. In most hypothesis tests, a test statistic is a particular data summary that is especially useful for computing the p-value and evaluating the hypothesis test. In the case of point estimates that are nearly normal, the test statistic is the Z score.

Test statistic

A *test statistic* is a special summary statistic that is especially useful for evaluating a hypothesis test or identifying the p-value. When a point estimate is nearly normal, we use the Z score of the point estimate as the test statistic. In later chapters we encounter point estimates where other test statistics are helpful.

In Chapter 5, we will apply our confidence interval and hypothesis testing framework to a variety of point estimates. In each case, we will require the sample size to be sufficiently large for the normal approximation to hold. In Chapter 6, we apply the ideas of confidence intervals and hypothesis testing to cases where the point estimate and test statistic is not necessarily normal. Such cases occur when the sample size is too small for the normal approximation, the standard error estimate may be poor, or the point estimate tends towards some other distribution.

4.6 Problem set

4.6.1 Variability in estimates

4.1 For each of the following situations, state whether the parameter of interest is the mean or the proportion. It may be helpful to examine whether individual responses are numeric or categorical. If numerical, parameter of interest is often the population mean, if categorical, the population proportion.

(a) In a survey, one hundred college students are asked how many hours per week they spend on the Internet.

(b) In a survey, one hundred college students are asked: "What percentage of the time you spend on the Internet is as part of your course work?"

(c) In a survey, one hundred college students are asked whether or not they cited information from Wikipedia on their papers.

(d) In a survey, one hundred college students are asked what percentage of their total weekly spending is on alcoholic beverages.

(e) In a sample of one hundred recent college graduates, it is found that 85 percent expect to get a job within one year of their graduation date.

4.2 A college counselor interested in finding out how many credits students typically take each semester randomly samples 100 students from the registrar database. The histogram below shows the distribution of the number of credits taken by these students and sample statistics for this distribution are also provided.

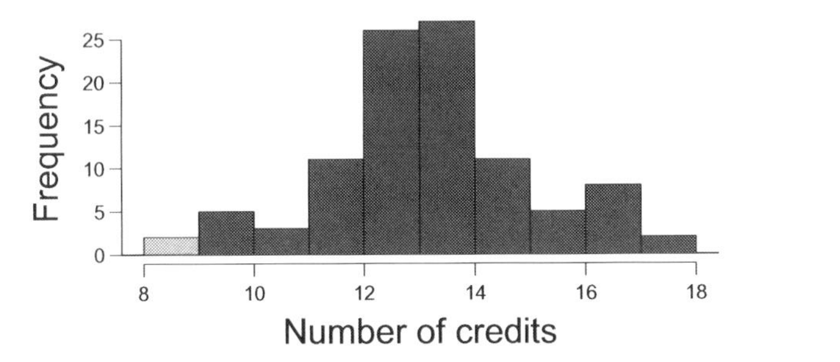

Min	8
Q1	13
Median	14
Mean	13.65
SD	1.91
Q3	15
Max	18

(a) What is the point estimate for the average number of credits taken per semester by students at this college? What about the median?

(b) What is the point estimate for the standard deviation of the number of credits taken per semester by students at this college? What about the IQR?

(c) Is a load of 16 credits unusually high for this college? What about 18 credits? Explain your reasoning.

4.3 The college counselor mentioned in Exercise 2 takes another random sample of 100 students and this time finds a sample mean of 14.02 and a standard deviation of 1.08. Should she be surprised that the sample statistics are slightly different for this sample? Explain your reasoning.

4.4 The mean number of credits taken by the random sample of college students in Exercise 2 is a point estimate for the mean number of credits taken by all students at that college. In Exercise 3 we saw that a new random sample gives a slightly different sample mean.

(a) What measure do we use to quantify the variability of a such an estimate?

(b) Based on the data given in Exercise 2, how much variability should the college counselor expect to see in the means number of credits of random samples of 100 students?

4.5 John is shopping for wireless routers and is overwhelmed by the number of options out there. In order to get a feel for the average price he takes a random sample of 75 routers and finds that the average price for this sample is $75 and the standard deviation is $25.

(a) Based on this information, how much variability should he expect to see in the mean prices of random samples of 75 wireless routers?

(b) A consumer reports website claims that the average price of routers is $80. John thinks that a difference of $5 in average prices is too high, that the information on the website is wrong. What do you think?

4.6.2 Confidence intervals

4.6 A large sized coffee at Starbucks is called a *venti* which means *twenty* in Italian because a venti cup is supposed to hold 20 ounces of coffee. Jonathan believes that his venti always has less than 20 ounces of coffee in it so he wants to test Starbucks' claim. He randomly chooses 50 Starbucks locations and gets a cup of venti coffee from each one. He then measures the amount of coffee in each cup and finds that the mean amount of coffee is 19.4 ounces and the standard deviation is 0.8 ounces.

(a) Are assumptions and conditions for inference satisfied?

(b) Calculate a 90% confidence interval for the true mean amount of coffee in Starbucks venti cups.

(c) Explain what this interval means in the context of this question.

(d) What does "90% confidence" mean?

(e) Do you think Starbucks' claim that a venti cup has 20 ounces of coffee is reasonable based on the confidence interval you just calculated?

(f) Would you expect a 95% confidence interval to be wider or narrower? Explain your reasoning.

4.7 In order to estimate the average lifespan of ball bearings produced by a certain machine, a factory worker randomly samples 75 ball bearings produced by that machine and measures their life span. He calculates an average lifespan of 6.85 working hours, with a standard deviation of 1.25 working hours. Below histogram shows the distribution of the lifespans of the ball bearings in this sample.

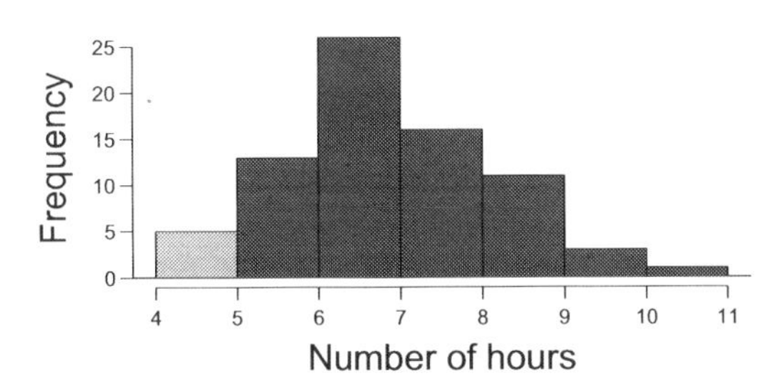

(a) Are assumptions and conditions for inference satisfied?

(b) Calculate a 98% confidence interval for the true average lifespan of ball bearings produced by this machine.

(c) Explain what this interval means in the context of this question.

(d) The manufacturer of the machine claims that the ball bearings produced by this machine last on average 7 hours. What do you think about this claim?

4.8 The distribution of weights of US pennies is approximately normal with a mean of 2.5 grams and a standard deviation of 0.4 grams.

(a) What is the probability that a randomly chosen penny weighs less than 2 grams?

(b) Describe the distribution of the mean weight of 30 randomly chosen pennies.

(c) What is the probability that the mean weight of 30 pennies is less than 2 grams?

(d) Sketch the two distributions (population and sampling) on the same scale.

(e) How would your answers in (a) and (b) change if the weights of pennies were not distributed normally?

4.9 A recent housing survey was conducted to determine the price of a typical home in Topanga, CA. The mean price of a house was roughly \$1.3 million with a standard deviation of \$300,000. There were no houses listed below \$600,000 but a few houses above \$3 million.

(a) Based on this information, do most houses in Topanga cost more or less than \$1.3 million?

(b) What is the probability that a randomly chosen house in Topanga costs more than \$1.5 million?

(c) What is the probability that the mean of 60 randomly chosen houses in Topanga is more than \$1.5 million?

(d) How would doubling the sample size affect the standard error of the mean?

4.10 If a higher confidence level means that we are more confident about the number we are reporting, why don't we always report a confidence interval with the highest confidence level possible?

4.11 In Section 4.3.2 a 95% confidence interval for the average age of runners in the 2009 Cherry Blossom Run was calculated to be (33.08, 37.36) based on a sample of 100 runners. How can we decrease the width of this interval without losing confidence?

4.12 A hospital administrator hoping to improve waiting times wishes to estimate the average waiting time at the emergency room at her hospital. She randomly selects 64 patients and determines the time (in minutes) between when they checked in to the ER until they were first seen by a doctor. A 95% confidence interval based on this sample is (128, 147). State if the following statements are true or false and explain your reasoning.

(a) This confidence interval is not valid since we do not know if the population distribution of the ER waiting times is nearly Normal.

(b) We are 95% confident that the average waiting time of these 64 patients is between 128 minutes and 147.

(c) We are 95% confident that the average waiting time of all patients at this hospital is between 128 minutes and 147.

(d) There is a 0.95 probability that the confidence interval includes the true average waiting time of all patients at this hospital.

(e) 95% of random samples have a sample mean between 128 and 147 minutes.

(f) A 99% confidence interval would be narrower than the 95% confidence interval since we need to be more sure of our estimate.

(g) The margin of error is 9.5 and the sample mean is 137.5.

(h) In order to decrease the margin of error of a 95% confidence interval to half of what it is now, we would need to double the sample size.

4.6.3 Hypothesis testing

4.13 Write the null and alternative hypotheses, in words and using symbols, for each of the following situations.

(a) New York is known as "the city that never sleeps". A random sample of 25 New Yorkers were asked how much sleep they get per night. Is there significant evidence to suggest that New Yorkers on average sleep less than 8 hrs a night?

(b) Since October 2008 chain restaurants in California are required to display calorie counts of each menu item. Prior to menus displaying calorie counts, the average calorie intake of a diner at a restaurant was 1900 calories. A nutritionist conducts a hypothesis test to see if there has been a significant decrease in the average calorie intake of a diners at this restaurant.

(c) Employers at a firm are worried about the effect of March Madness on employees' productivity. They estimate that on a regular business day employees spend on average 15 minutes of company time checking personal email, making personal phone calls etc., i.e. not working. They want to test if productivity significantly decreases during March Madness.

(d) Based on the performance of examinees who took the GRE exam between July 1, 2004 and June 30, 2007, the average Verbal Reasoning score was calculated to be 462.[9] Is there evidence to suggest that the average GRE Verbal Reasoning score has changed since 2007?

4.14 Exercise 9 presents the results of a recent survey showing that the mean price of a house in Topanga, CA is \$1.3 million. A prospective homeowner does not believe that this figure is an overestimation and decides to collect his own sample and conduct a hypothesis test. Below is how he set up his hypotheses. Indicate any errors you see.

$$H_0 : \bar{x} = \$1.3 \text{ million}$$
$$H_A : \bar{x} > \$1.3 \text{ million}$$

4.15 Exercise 12 provides a 95% confidence interval for the mean waiting time at an ER. A local newspaper claims that the average waiting time at this ER exceeds 3 hours. What do you think of their claim?

4.16 Based on the confidence interval provided in Exercise 12, the hospital administrator found that 2.2 hours was not implausible based on the 95% confidence interval. Would she arrive at the same conclusion using a 99% interval? Why or why not?

4.17 A patient is diagnosed with Fibromyalgia and prescribed anti-depressants. Being the skeptic that she is, she doesn't initially believe that anti-depressants will treat her symptoms. However after a couple months of being on the medication she decides that the anti-depressants are working since she feels like her symptoms are in fact getting better.

(a) Write the hypotheses in words.

(b) What is a Type I error in this context?

(c) What is a Type II error in this context?

(d) How would these errors affect the patient?

4.18 A food safety inspector is called upon to inspect a restaurant that received a few complaints from diners who claimed that food safety and sanitation regulations were not being met at this restaurant. The food safety inspector approaches this task as a hypothesis test, examines whether the data provides convincing evidence that the regulations are not met. If he does, the restaurant gets shut down.

(a) Write the hypotheses in words.

(b) What is a Type I error in this context?

(c) What is a Type II error in this context?

(d) Which error may be more problematic for the restaurant owner?

(e) Which error may be more problematic for the diners?

4.19 A 2000 survey showed that on average 25% of the time college students spent on the Internet was for coursework. A decade later in 2010 a new survey was given to 50 randomly sampled college students. The responses showed that on average 10% of the time college students spent on the Internet was for coursework, with a standard deviation of 3%. Is there evidence to suggest that the percentage of time college students spend on the Internet for coursework has decreased over the last decade?

(a) Are assumptions and conditions for inference met?

(b) Perform an appropriate hypothesis test and state your conclusion.

(c) Interpret the p-value in context.

(d) What type of an error might you have made?

4.20 The hospital administrator mentioned in Exercise 12 randomly selected 64 patients and determines the time (in minutes) between when they checked in to the ER until they were first seen by a doctor. The average time is 137.5 minutes and the standard deviation is 37 minutes. She is getting grief from her supervisor that the wait times in the ER has increased greatly from last year's average of 128 minutes however she claims that the increase is not statistically significant.

(a) Are assumptions and conditions for inference met?

(b) Using a significance level of $\alpha = 0.05$, is the increase in wait times statistically significant?

(c) Would the conclusion of the hypothesis test change if the significance level was changed to $\alpha = 0.01$?

4.21 Exercise 7 provides information on the average lifespan and standard deviation of 75 randomly sampled ball bearings produced by a certain machine. We are also told that the manufacturer of the machine claims that the ball bearings produced by this machine last 7 hours on average. Conduct a formal hypothesis test to test this claim.

4.22 The nutrition label on a bag of potato chips says that a one ounce (28 gram) serving of potato chips has 130 calories and contains ten grams of fat, with three grams of saturated fat. A random sample of 55 bags yielded a sample mean of 138 calories with a standard deviation of 17 calories. Is there evidence to suggest that the nutrition label does not provide an accurate measure of calories in the bags of potato chips?

4.6.4 Examining the Central Limit Theorem

4.23 An iPod has 3,000 songs. The histogram below shows the distribution of the lengths of these songs. It is also given that the mean length is 3.45 minutes and the standard deviation is 1.63 minutes.

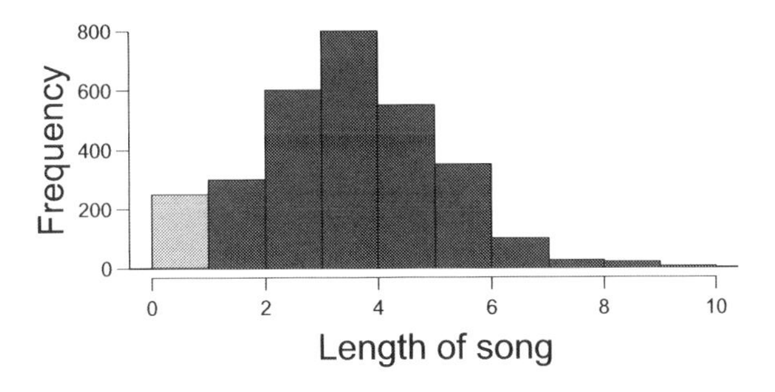

(a) Calculate the probability that a randomly selected song lasts more than 5 minutes.

(b) You are about to go for an hour run and you make a random playlist of 15 songs. What is the probability that your playlist lasts for the entire duration of your run?

(c) You are about to take a trip to visit your parents and the drive is 6 hours. You make a random playlist of 100 songs. What is the probability that your playlist lasts the entire drive?

4.24 Suppose that the area that can be painted using a single can of spray paint is slightly variable and follows a normal distribution with a mean of 25 square feet and a standard deviation of 3 square feet.

(a) What is the probability that the area covered by a can of spray paint is more than 27 square feet?

(b) What is the probability that the area covered by 20 cans of spray paint is at least 540 square feet?

(c) How would your answers in (a) and (b) change if the area covered by a can of spray paint is not distributed normally?

4.6.5 A framework for inference

4.25 A car insurance company advertises that customers switching to their insurance save \$432 on average on their yearly premiums. A market researcher who thinks this is an overestimate and false advertisement randomly samples 82 customers who recently switched to this insurance and finds an average savings of \$395, with a standard deviation of \$102.

(a) Are assumptions and conditions for inference satisfied?

(b) Perform a hypothesis test and state your conclusion.

(c) Do you agree with the market researcher that the amount of savings advertised is an overestimate? Explain your reasoning.

(d) Calculate a 90% confidence interval for the average amount of savings of all customers who switch their insurance.

(e) Do your results from the hypothesis test and the confidence interval agree? Explain.

4.26 A restaurant owner is considering extending the happy hour at his restaurant since he believes that an extended happy hour may increase the revenue per customer. He estimates that currently average revenue per customer is \$18 during happy hour. He runs extended happy hour for a week and based on a random sample of 70 customers the average revenue is \$19.25 with a standard deviation of \$3.02.

(a) Are assumptions and conditions for inference satisfied?

(b) Perform a hypothesis test. Based on the result of the hypothesis test, is there significant evidence to suggest that the revenue per customer has increased when happy hour was extended by an hour?

(c) Calculate a 90% confidence interval for the average amount of savings of all customers who switch their insurance.

(d) Do your results from the hypothesis test and the confidence interval agree? Explain.

(e) If your hypothesis test and confidence interval suggest a significant increase in revenue per customer, why might you still not recommend that the restaurant owner extend the happy hour based on this criterion? What may be a better measure to consider?

Chapter 5

Large sample inference

Chapter 4 introduced a framework of statistical inference through estimation of the population mean μ. Chapter 5 explores statistical inference for three slightly more interesting cases: (5.1, 5.2) differences of two population means, (5.3) a single population proportion and (5.4) differences of two population proportions. In each application, we verify conditions that ensure the point estimates are nearly normal and then apply the general framework from Section 4.5.

In Chapter 6 we will extend our reach to smaller samples. In such cases, point estimates may not be nearly normal, and we will be required to identify the distribution of the point estimate with greater detail.

5.1 Paired data

Are textbooks actually cheaper online? Here we compare the price of textbooks at UCLA's bookstore and prices at Amazon.com. Seventy-three UCLA courses were randomly sampled in Spring 2010, representing less than 10% of all UCLA courses[1]. A portion of this `textbooks` data set is shown in Table 5.1.

5.1.1 Paired observations and samples

Each textbook has two corresponding prices in the data set: one for the UCLA bookstore and one for Amazon. In this way, each textbook price from the UCLA bookstore has a natural correspondence with a textbook price from Amazon (if

[1] When a class had multiple books, only the most expensive text was considered.

	deptAbbr	course	uclaNew	amazNew	diff
1	Am Ind	C170	27.67	27.95	-0.28
2	Anthro	9	40.59	31.14	9.45
⋮	⋮	⋮	⋮	⋮	⋮
73	Wom Std	285	27.70	18.22	9.48

Table 5.1: Three cases of the `textbooks` data set.

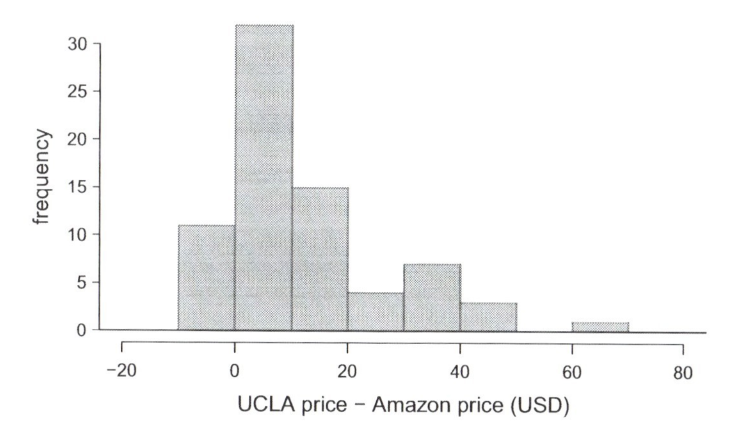

Figure 5.2: Histogram of the difference in price for each book sampled.

they correspond to the same book). When two sets of observations have thi special correspondence, they are said to be **paired**.

> **Paired data**
> Two sets of observations are *paired* if each observation in one set has a special correspondence with exactly one observation in the other data set.

To analyze paired data, it is often useful to look at the difference in each pair of observations. In the `textbook` data set, we look at the difference in prices, which is represented as the `diff` variable in the `textbooks` data. Here the differences are taken as

$$\text{UCLA price} - \text{Amazon price}$$

for each book, which can be verified by examining the three entries in Table 5.1. It is important that we always subtract using a consistent order; here Amazon prices are always subtracted from UCLA prices. A histogram of these differences is shown in Figure 5.2. Using differences between paired observations is a common and useful way to analyze paired data.

⊙ **Exercise 5.1** The first difference shown in Table 5.1 is computed as $27.67 - 27.95 =$

−0.28. Verify the other two differences in the table are computed correctly.

5.1.2 Inference for paired data

To analyze a paired data set, we use the exact same tools we developed in Chapter 4 and we simply apply them to the differences in the paired observations.

n_{diff}	$\bar{x}_{diff}$	s_{diff}
73	12.76	14.26

Table 5.3: Summary statistics for the diff variable. There were 73 books, so there are 73 differences.

● **Example 5.1** Set up and implement a hypothesis test to determine whether Amazon's average textbook price is different than the UCLA bookstore's average textbook price.

There are two scenarios: there is no difference or there is a difference in average prices. The *no difference* scenario is always the null hypothesis:

H_0: $\mu_{diff} = 0$. There is no difference in the average textbook price. The notation μ_{diff} is used as a notational reminder that we should only work with the difference in prices.

H_A: $\mu_{diff} \neq 0$. There is a difference in average prices.

Can the normal model be used to describe the sampling distribution of $\bar{x}_{diff}$? To check, the conditions must be satisfied for the diff data. The observations are from a simple random sample from less than 10% of the all books sold at the bookstore, so independence is reasonable. There is skew in the differences (Figure 5.2), however, it is not extreme. There are also more than 50 differences. Thus, we can conclude the sampling distribution of $\bar{x}_{diff}$ is nearly normal.

We can compute the standard error associated with $\bar{x}_{diff}$ using the standard deviation of the differences and the number of differences:

$$SE_{\bar{x}_{diff}} = \frac{s_{diff}}{\sqrt{n_{diff}}} = \frac{14.26}{\sqrt{73}} = 1.67$$

To visualize the p-value, the sampling distribution of $\bar{x}_{diff}$ is drawn under the condition as though H_0 was true, which is shown in Figure 5.4. The p-value is represented by the two (very) small tails.

To find the tail areas, we compute the test statistic, which is the Z score of $\bar{x}_{diff}$ under the condition $\mu_{diff} = 0$:

$$Z = \frac{\bar{x}_{diff} - 0}{SE_{x_{diff}}} = \frac{12.76 - 0}{1.67} = 7.59$$

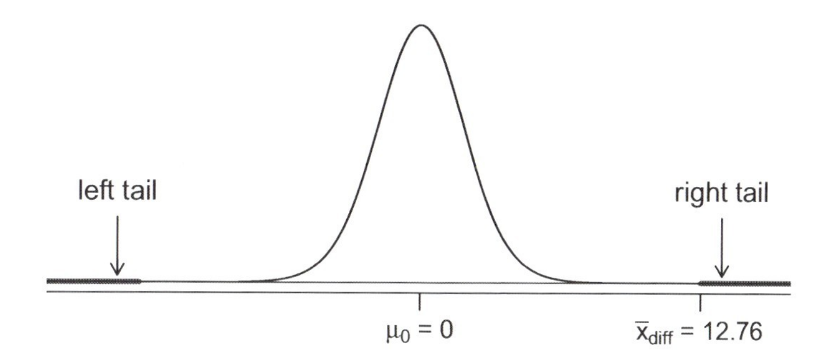

Figure 5.4: Histogram of the difference in price for each book sampled.

This Z score is so large it isn't even in the table, which ensures the single tail area will be 0.0002 or smaller. Since the p-value is both tails, the p-value can be estimated as

$$\text{p-value} = 2 * (\text{one tail area}) = 2 * 0.0002 = 0.0004$$

Because this is less than 0.05, we reject the null hypothesis. We have found convincing evidence that Amazon is, on average, cheaper than the UCLA bookstore.

⊙ **Exercise 5.2** Create a 95% confidence interval for the average price difference between books at the UCLA bookstore and books on Amazon. [2]

5.2 Difference of two means

In this section we consider a difference in two population means, $\mu_1 - \mu_2$, under the condition that the data are not paired. The methods are similar in theory but different in the details. Just as with a single sample, we identify conditions to ensure a point estimate of the difference, $\bar{x}_1 - \bar{x}_2$, is nearly normal. Next we introduce a formula for the standard error, which allows us to apply our general tools from Section 4.5.

We apply these methods to two examples: participants in the 2009 Cherry Blossom Run and newborn infants. This section is motivated by questions like "Is there convincing evidence that newborns from mothers who smoke have a different average birth weight than newborns from mothers who don't smoke?"

[2]Conditions have already verified and the standard error computed in Example 5.1. To find the interval, identify z^* (1.96 for 95% confidence) and plug it, the point estimate, and the standard error into the general confidence interval formula:

$$\text{point estimate} \pm z^* SE \quad \rightarrow \quad 12.76 \pm (1.96)(1.67) \quad \rightarrow \quad (9.49, 16.03)$$

We are 95% confident that Amazon is, on average, between \$9.49 and \$16.03 cheaper than the UCLA bookstore for each UCLA course book.

	men	women
$\bar{x}$	88.08	96.28
s	15.74	13.66
n	100	80

Table 5.5: Summary statistics for the run time of 180 participants in the 2009 Cherry Blossom Run.

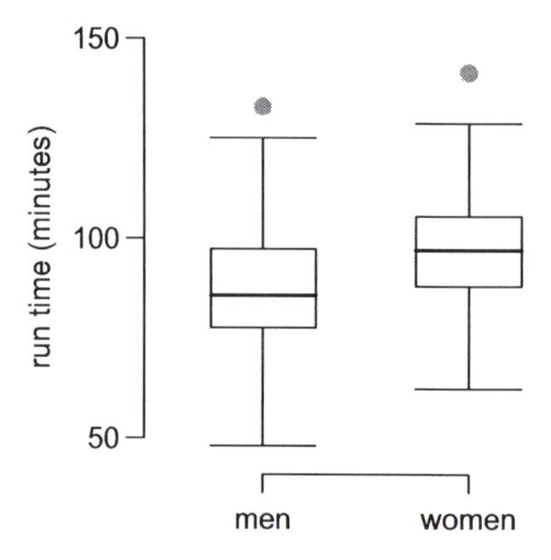

Figure 5.6: Side-by-side box plots for the sample of 2009 Cherry Blossom Run participants.

5.2.1 Point estimates and standard errors for differences of means

We would like to estimate the average difference in run times for men and women using a random sample of 100 men and 80 women from the `run10` population. Table 5.5 presents relevant summary statistics, and box plots of each sample are shown in Figure 5.6.

The two samples are independent of one-another, so the data are not paired. Instead a point estimate of the difference in average 10 mile times for men and women, $\mu_w - \mu_m$, can be found using the two sample means:

$$\bar{x}_w - \bar{x}_m = 8.20$$

Because we are examining two simple random samples from less than 10% of the population, each sample contains at least 50 observations, and neither distribution is strongly skewed, we can safely conclude the two sample means are each nearly normal. Finally, because each sample is independent of the other (e.g. the data are not paired), we can conclude that the difference in sample means is nearly normal[3].

[3]Probability theory guarantees that the difference of two independent normal random variables is also normal. Because each sample mean is nearly normal and observations in the samples are independent, we are assured the difference is also nearly normal.

Conditions for normality of $\bar{x}_1 - \bar{x}_2$
If the sample means, $\bar{x}_1$ and $\bar{x}_2$, each meet the criteria for being nearly normal and the observations in the two samples are independent, then the difference in sample means, $\bar{x}_1 - \bar{x}_2$, is nearly normal.

We can quantify the variability in the point estimate, $\bar{x}_w - \bar{x}_m$, using the following formula for its standard error:

$$SE_{\bar{x}_w - \bar{x}_m} = \sqrt{\frac{\sigma_w^2}{n_w} + \frac{\sigma_m^2}{n_m}}$$

We usually estimate this standard error using standard deviation estimates based on the samples:

$$\begin{aligned} SE_{\bar{x}_w - \bar{x}_m} &= \sqrt{\frac{\sigma_w^2}{n_w} + \frac{\sigma_m^2}{n_m}} \\ &\approx \sqrt{\frac{s_w^2}{n_w} + \frac{s_m^2}{n_m}} = \sqrt{\frac{13.7^2}{80} + \frac{15.7^2}{100}} = 2.19 \end{aligned}$$

Because each sample has at least 50 observations ($n_w = 80$ and $n_m = 100$), this substitution using the sample standard deviation tends to be very good.

Distribution of a difference of sample means
The sample difference of two means, $\bar{x}_1 - \bar{x}_2$, is nearly normal with mean $\mu_1 - \mu_2$ and estimated standard error

$$SE_{\bar{x}_1 - \bar{x}_2} = \sqrt{\frac{s_1^2}{n_1} + \frac{s_2^2}{n_2}} \qquad (5.1)$$

when each sample mean is nearly normal and all observations are independent.

5.2.2 Confidence interval for the difference

When the conditions are met for $\bar{x}_1 - \bar{x}_2$ to be nearly normal, we can construct a 95% confidence interval for the difference in two means using the framework built in Section 4.5. Here a point estimate, $\bar{x}_w - \bar{x}_m = 8.20$, is associated with a normal model with standard error 2.19. Using this information, the general confidence interval formula may be applied to try to capture the true difference in means:

$$pointEstimate \pm z^* SE \quad \rightarrow \quad 8.20 \pm 1.96 * 2.19 \quad \rightarrow \quad (3.91, 12.49)$$

Based on the samples, we are 95% confident that men ran, on average, between 3.91 and 12.49 minutes faster than women in the 2009 Cherry Blossom Run.

diff. of 2 means

	fAge	mAge	weeks	weight	sexBaby	smoke
1	NA	13	37	5.00	female	nonsmoker
2	NA	14	36	5.88	female	nonsmoker
3	19	15	41	8.13	male	smoker
⋮	⋮	⋮	⋮	⋮	⋮	
150	45	50	36	9.25	female	nonsmoker

Table 5.7: Four cases from the `babySmoke` data set. Observations listed as "NA" listings mean that particular piece of data is missing.

⊙ **Exercise 5.3** What does 95% confidence mean? [4].

⊙ **Exercise 5.4** We may be interested in a different confidence level. Construct the 99% confidence interval for the true difference in average run times based on the sample data. Hint in the footnote[5].

5.2.3 Hypothesis test based on a difference in means

A data set called `babySmoke` represents a random sample of 150 cases of mothers and their newborns in North Carolina over a year. Four cases from this data set are represented in Table 5.7. We are particularly interested in two variables: `weight` and `smoke`. The `weight` variable represents the weights of the newborns and the `smoke` variable describes which mothers smoked during pregnancy. We would like to know, is there convincing evidence that newborns from mothers who smoke have a different average birth weight than newborns from mothers who don't smoke? We will answer this question using a hypothesis test. The smoking group includes 50 cases and the nonsmoking group contains 100 cases, represented in Figure 5.8.

⊙ **Exercise 5.5** Set up appropriate hypotheses to evaluate whether there is a relationship between a mother smoking and average birth weight. Write out the hypotheses both in plain language and in statistical notation. Answer in the footnote[6].

Summary statistics are shown for each sample in Table 5.9. Because each sample is simple random and consists of less than 10% of all such cases, the observations are independent. Additionally, each sample size is at least 50 and neither

---

[4] If we were to collected many such samples and create 95% confidence intervals for each, then about 95% of these intervals would contain the population difference in average run times for men and women, $\mu_w - \mu_m$.

[5] The only thing that changes is z^*: we use $z^* = 2.58$ for a 99% confidence level. If the selection of z^* is confusing, see Section 4.2.4 for an explanation.

[6] The null hypothesis represents the case of no difference between the groups. H_0: there is no difference in average birth weight for newborns from mothers who did and did not smoke. In statistical notation: $\mu_n - \mu_s = 0$, where μ_n represents non-smoking mothers and μ_s represents mothers who smoked. H_A: there is some difference in average newborn weights from mothers who did and did not smoke ($\mu_n - \mu_s \neq 0$).

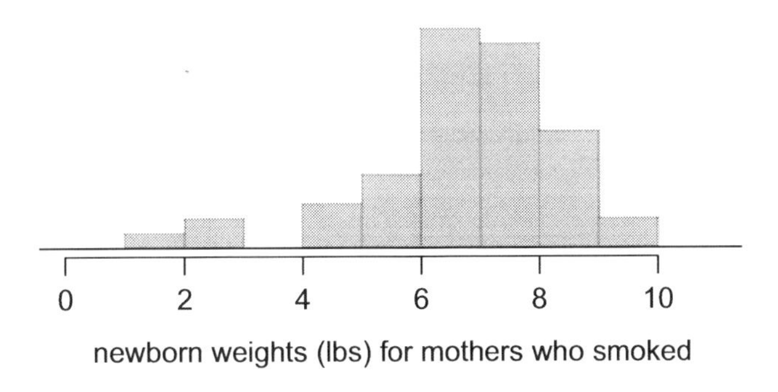

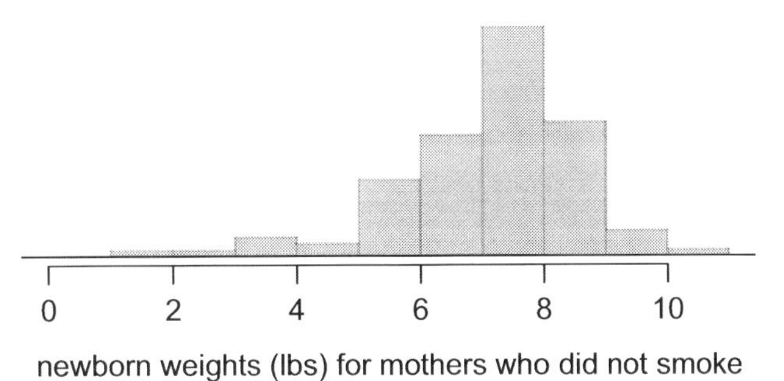

Figure 5.8: The top panel represents birth weights for infants whose mothers smoked. The bottom panel represents the birth weights for infants whose mothers who did not smoke.

sample distribution is strongly skewed (see Figure 5.8), so both sample means can be associated with the normal model.

	smoker	nonsmoker
mean	6.78	7.18
st. dev.	1.43	1.60
samp. size	50	100

Table 5.9: Summary statistics for the `babySmoke` data set.

⊙ **Exercise 5.6** (a) What is the point estimate of the population difference, $\mu_n - \mu_s$? (b) What is the distribution of this point estimate? (The two samples are independent of each other.) (c) Compute the standard error of the point estimate from part (a). [7].

● **Example 5.2** If the null hypothesis was true, what would be the expected value of the point estimate from Exercise 5.6? And the standard deviation of this estimate? Draw a picture to represent the p-value.

If the null hypothesis was true, then we expect to see a difference near 0. The standard error corresponds to the standard deviation of the point estimate: 0.26. To depict the p-value, we draw the distribution of the point estimate as though H_0 was true:

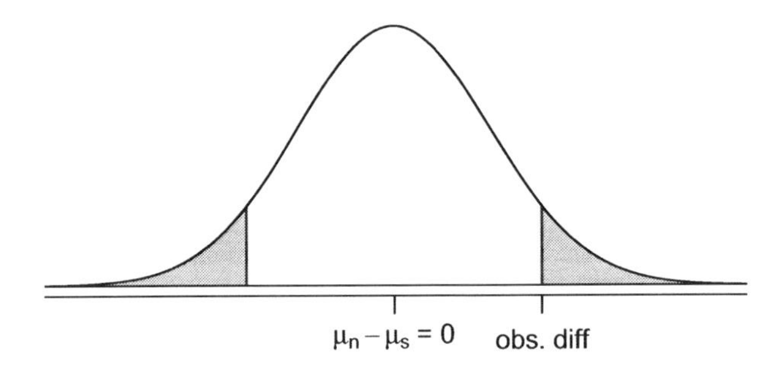

Both tails are shaded because it is a two-sided test.

● **Example 5.3** Compute the p-value of the hypothesis test using the figure in Example 5.2 and evaluate the hypotheses using a significance level of $\alpha = 0.05$.

[7](a) The difference in sample means is an appropriate point estimate: $\bar{x}_n - \bar{x}_s = 0.40$. (b) Because the samples are independent and each sample mean is nearly normal, their difference is also nearly normal. (c) The standard error of the estimate can be estimated using Equation (5.1):

$$SE = \sqrt{\frac{\sigma_n^2}{n_n} + \frac{\sigma_s^2}{n_s}} \approx \sqrt{\frac{s_n^2}{n_n} + \frac{s_s^2}{n_s}} = \sqrt{\frac{1.60^2}{100} + \frac{1.43^2}{50}} = 0.26$$

The sample standard deviations can be used because our conditions are met for the normal model.

Since the point estimate is nearly normal, we can find the upper tail using the Z score and normal probability table:

$$Z = \frac{0.40 - 0}{0.26} = 1.54 \quad \rightarrow \quad \text{upper tail } = 1 - 0.938 = 0.062$$

Because this is a two-sided test and we want the area of both tails, we double this single tail to get the p-value: 0.124. This p-value is larger than the significance value, 0.05, so we fail to reject the null hypothesis. There is insufficient evidence to say there is a difference in average birth weight of newborns from mothers who did smoke during pregnancy and newborns from mothers who did not smoke during pregnancy.

⊙ **Exercise 5.7** Does the conclusion to Example 5.3 mean that smoking and average birth weight are unrelated? [8]

⊙ **Exercise 5.8** If we actually did make a Type 2 Error and there is a difference, what might we have done differently in data collection to be more likely to detect such a difference? [9]

5.2.4 Summary for inference of the difference of two means

When considering the difference of two means, there are two cases: the data observed are paired or not. The first case was treated in Section 5.1, where the one-sample methods were applied to the differences from the paired observations. We examined the second and more complex scenario in this section.

When applying the normal model to the point estimate, $\bar{x}_1 - \bar{x}_2$, when the data are not paired, it is important to first verify conditions before applying the inference framework using the normal model. First, each sample mean must meet the conditions for normality; these conditions are described in Chapter 4. Secondly, all observations in the collected samples are independent. When these conditions are satisfied, the framework from Section 4.5 may be applied.

For example, a general confidence interval takes the following form:

$$\text{point estimate} \pm z^* SE$$

When estimating $\mu_1 - \mu_2$, the point estimate would be the difference in sample means, the value z^* would correspond to the confidence level, and the standard error would be computed from Equation (5.1) on page 175. While the point estimate and standard error formulas change a little, the general framework for a confidence interval stays the same. This holds true also for hypothesis tests for differences of means.

[8] Absolutely not. It is possible that there is some difference but we did not detect it. If this is the case, we made a Type 2 Error.

[9] We could have collected larger samples. If the sample size is larger, we tend to have a better shot at finding a difference if one exists.

In a hypothesis test, we proceed just as before but again use different point estimates and standard errors when working with the difference of two means. The test statistic represented by the Z score may be computed as

$$Z = \frac{\text{point estimate} - \text{null value}}{SE}$$

When assessing the difference in two means, the point estimate takes the form $\bar{x}_1 - \bar{x}_2$, and the standard error again takes the form of Equation (5.1) on page 175. Finally, the null value is the difference in sample means under the null hypothesis. Just as in Chapter 4, the test statistic Z is used to identify the p-value.

5.2.5 Examining the standard error formula

The formula for standard error for the difference in two means has structure reflecting that of the single sample standard error. Recall that the standard error of a single mean, $\bar{x}_1$, can be approximated by

$$SE_{\bar{x}_1} = \frac{s_1}{\sqrt{n_1}}$$

where s_1 and n_1 represent the sample standard deviation and sample size.

The standard error of the difference of two sample means can be constructed from the standard errors of the separate sample means:

$$SE_{\bar{x}_1 - \bar{x}_2} = \sqrt{SE^2_{\bar{x}_1} + SE^2_{\bar{x}_2}} = \sqrt{\frac{s_1^2}{n_1} + \frac{s_2^2}{n_2}} \tag{5.2}$$

This special relationship follows from some probability theory; see Section 2.6 for details. [Authors' note: Section 2.6 will be released in the First Edition in 2011.]

⊙ **Exercise 5.9** Prerequisite: Section 2.6. We can rewrite Equation (5.2) in a different way:

$$SE^2_{\bar{x}_1 - \bar{x}_2} = SE^2_{\bar{x}_1} + SE^2_{\bar{x}_2}$$

Explain where this formula comes from using the ideas of probability theory. Hint in the footnote[10].

5.3 Single population proportion

According to a poll taken by CNN/Opinion Research Corporation in February 2010, only about 26% of the American public trust the federal government most of the time[11]. This poll included responses of 1,023 Americans.

We will find that sample proportions, like sample means, tend to be nearly normal when the sample size is sufficiently large.

---

[10] The standard error squared represents the variance of the estimate. If X and Y are two random variables with variances σ_x^2 and σ_y^2, what is the variance of $X - Y$?

[11] http://www.cnn.com/2010/POLITICS/02/23/poll.government.trust/index.html

5.3.1 Identifying when a sample proportion is nearly normal

A sample proportion can be described as a sample mean. If we represent each "success" as a 1 and each "failure" as a 0, then the sample proportion is equal to the mean of these numerical outcomes:

$$\hat{p} = \frac{0 + 0 + 1 + \cdots + 0}{1015} = 0.26$$

The distribution of $\hat{p}$ is nearly normal when the distribution of 0's and 1's is not too strongly skewed or the sample size is sufficiently large. The most common guidelines for sample size and skew is to ensure that we expect to observe a minimum number of successes and failures, typically at least 10 of each.

Conditions for $\hat{p}$ being nearly normal
A sample proportion $\hat{p}$, taken from a sample of size n from a population with a true proportion p, is nearly normal when

1. the sample observations are independent and
2. we expected to see at least 10 successes and 10 failures in our sample, i.e. $np \geq 10$ and $n(1-p) \geq 10$. This is called the **success-failure condition.**

If these conditions are met, then $\hat{p}$ is nearly normal with mean p and standard error

$$SE_{\hat{p}} = \sqrt{\frac{p(1-p)}{n}} \qquad (5.3)$$

Typically we do not know the true proportion, p, so must substitute some value to check conditions and to estimate the standard error. For confidence intervals, usually $\hat{p}$ is used to check the success-failure condition and compute the standard error. For hypothesis tests, typically the null value – that is, the proportion claimed in the null hypothesis – is used in place of p. Examples are presented for each of these cases in Sections 5.3.2 and 5.3.3.

TIP: Checking independence of observations
If our data comes from a simple random sample and consists of less than 10% of the population, then the independence assumption is reasonable. Alternatively, if the data comes from a random process, we must evaluate the independence condition more carefully.

5.3.2 Confidence intervals for a proportion

We may want a confidence interval for the proportion of Americans who do not trust federal officials. Our estimate is $\hat{p} = 0.26$ based on the sample of size $n = 1023$. To use the general confidence interval formula from Section 4.5, we check conditions to verify $\hat{p}$ is nearly normal and also identify the standard error of this estimate.

The sample is simple and random, and 1023 people is certainly less than 10% of the total US population, so independence is confirmed. The sample size must also be sufficiently large, which is checked via the success-failure condition: there were approximately $1023 * \hat{p} = 266$ "successes" and $1023(1 - \hat{p}) = 757$ "failures" in the sample, both easily greater than 10.

With the conditions met, we are assured that $\hat{p}$ is nearly normal. Next, a standard error for $\hat{p}$ is needed, and then the usual method to construct a confidence interval may be used.

⊙ **Exercise 5.10** Estimate the standard error of $\hat{p} = 0.26$ using Equation (5.3). Because p is unknown and the standard error is for a confidence interval, use $\hat{p}$ in place of p. [12]

● **Example 5.4** Construct a 95% confidence interval for p, the proportion of Americans who trust federal officials most of the time.

Using the standard error estimate from Exercise 5.10, the point estimate 0.26, and $z^* = 1.96$ for a 95% confidence interval, the general confidence interval formula from Section 4.5 may be used:

$$\text{point estimate } \pm z^* SE \quad \rightarrow \quad 0.26 \pm 1.96 * 0.014 \quad \rightarrow \quad (0.233, 0.287)$$

We are 95% confident that the true proportion of Americans who trusted federal officials most of the time (in February 2010) is between 0.233 and 0.287. If the proportion has not changed, not many Americans are very trusting of the federal government.

Constructing a confidence interval for a proportion

- Verify the observations are independent and also verify the success-failure condition using $\hat{p}$.
- If the conditions are met, $\hat{p}$ may be well-approximated by the normal model.
- Construct the standard error using $\hat{p}$ in place of p and apply the general confidence interval formula.

5.3.3 Hypothesis testing for a proportion

To apply the normal model tools in the context of a hypothesis test for a proportion, the independence and success-failure conditions must be satisfied. In a hypothesis test, the success-failure condition is checked using the null proportion: we would verify np_0 and $n(1 - p_0)$ are at least 10, where p_0 is the null value.

[12] The computed result is $SE = 0.014$.

⊙ **Exercise 5.11** Deborah Toohey is running for Congress, and her campaign manager claims she has more than 50% support from the district's electorate. Set up a hypothesis test to evaluate this (one-sided) claim. [13]

● **Example 5.5** A newspaper collects a simple random sample of 500 likely voters in the district and finds Toohey's support at 52%. Does this provide convincing evidence for the claim of Toohey's manager at the 5% significance level?

Because the sample is simple random and includes less than 10% of the population, the observations are independent. In a one-proportion hypothesis test, the success-failure condition is checked using the null proportion, $p_0 = 0.5$: $np_0 = n(1 - p_0) = 500 * 0.5 = 250 > 10$. With these conditions verified, the normal model may be applied to $\hat{p}$.

Next the standard error can be computed. Again the null value is used because this is a hypothesis test for a single proportion.

$$SE = \sqrt{\frac{p_0(1 - p_0)}{n}} = \sqrt{\frac{0.5 * (1 - 0.5)}{500}} = 0.022$$

A picture of the normal model is shown in Figure 5.10 with the p-values represented. Based on the normal model, the test statistic can be computed as the Z score of the point estimate:

$$Z = \frac{\text{point estimate} - \text{null value}}{SE} = \frac{0.52 - 0.50}{0.022} = 0.89$$

The upper tail area, representing the p-value, is 0.186. Because the p-value is larger than 0.05, we failed to reject the null hypothesis. We did not find convincing evidence to support the campaign manager's claim.

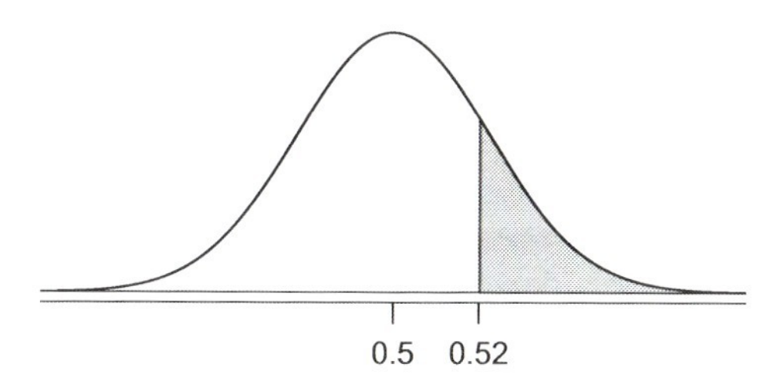

Figure 5.10: Distribution of the sample proportion if the null hypothesis was true. The p-value for the test is shaded.

[13]Is there convincing evidence that the campaign manager is correct? $H_0 : p = 0.50$, $H_A : p > 0.50$.

Hypothesis test for a proportion
Setup hypotheses and verify the conditions using the null value, p_0, to ensure $\hat{p}$ is nearly normal under H_0. If the conditions hold, construct the standard error, again using p_0, and depict the p-value in a drawing. Lastly, compute the p-value and evaluate the hypotheses.

5.4 Difference of two proportions

We would like to make conclusions about the difference in two population proportions: $p_1 - p_2$. We consider three examples. In the first, we construct a confidence interval for the difference in support for healthcare reform between Democrats and Republicans. In the second application, a company weighs whether they should switch to a higher quality parts manufacturer. In the last example, we examine the cancer risk to dogs from the use of yard herbicides.

In our investigations, we first identify a reasonable point estimate of $p_1 - p_2$ based on sample. You may have already guessed its form: $\hat{p}_1 - \hat{p}_2$. Next, we verify that the point estimate follows the normal model by checking conditions. Finally, we compute the estimates standard error and apply our inference framework, just as we have done in many cases before.

5.4.1 Distribution of the difference of two proportions

We check two conditions before applying the normal model to $\hat{p}_1 - \hat{p}_2$. First, each sample proportion must be nearly normal. Secondly, the samples must be independent. Under these two conditions, the sampling distribution of $\hat{p}$ may be well approximated using the normal model.

Conditions for $\hat{p}_1 - \hat{p}_2$ following a normal model
The difference in two sample proportions, $\hat{p}_1 - \hat{p}_2$, tends to follow a normal model if

- both proportions separately follow a normal model and
- the samples are independent.

The standard error of the difference in sample proportions is

$$SE_{\hat{p}_1 - \hat{p}_2} = \sqrt{SE_{\hat{p}_1}^2 + SE_{\hat{p}_2}^2} = \sqrt{\frac{p_1(1-p_1)}{n_1} + \frac{p_2(1-p_2)}{n_2}} \qquad (5.4)$$

where p_1 and p_2 represent the population proportions, and n_1 and n_2 represent the sample sizes.

For the difference in two means, the standard error formula took the following form:

$$SE_{\bar{x}_1-\bar{x}_2} = \sqrt{SE_{\bar{x}_1}^2 + SE_{\bar{x}_2}^2}$$

The standard error for the difference in two proportions takes a similar form. The reasons behind this similarity are rooted in the probability theory presented in Section 2.6. [Authors' note: Section 2.6 will be included in the 2011 release.]

5.4.2 Intervals and tests for $p_1 - p_2$

Just as with the case of a single proportion, the sample proportions are used to verify the success-failure condition and also compute standard error when constructing confidence intervals.

● **Example 5.6** One way to measure how bipartisan an issue is would be to compare the issue's support by Democrats and Republicans. In a January 2010 Gallup poll, 82% of Democrats supported a vote on healthcare in 2010 while only 20% of Republicans support such action[14]. This support is summarized in Figure 5.11. The sample sizes were 325 and 172 for Democrats and Republicans, respectively. Create and interpret a 90% confidence interval of the difference in support between the two parties.

First the conditions must be verified. Because this is a random sample from less than 10% of the population, the observations are independent, both within the samples and between the samples. The success-failure condition also holds using the sample proportions (for each sample). Because our conditions are met, the normal model can be used for the point estimate of the difference in support:

$$\hat{p}_D - \hat{p}_R = 0.82 - 0.20 = 0.62$$

The standard error may be computed using Equation (5.4) with each sample proportion:

$$SE \approx \sqrt{\frac{0.82(1-0.82)}{325} + \frac{0.20(1-0.20)}{172}} = 0.037$$

For a 90% confidence interval, we use $z^* = 1.65$:

$$\text{point estimate} \pm z^*SE \quad \rightarrow \quad 0.62 \pm 1.65 * 0.037 \quad \rightarrow \quad (0.56, 0.68)$$

We are 90% confident that the difference in support for healthcare action between the two parties is between 56% and 68%. Healthcare is a very partisan issue, which may not be a surprise to anyone who followed the health care debate in early 2010.

⊙ **Exercise 5.12** A remote control car company is considering a new manufacturer for wheel gears. The new manufacturer would be more expensive but their higher quality gears are more reliable, resulting in happier customers and fewer warranty claims.

[14] http://www.gallup.com/poll/125030/Healthcare-Bill-Support-Ticks-Up-Public-Divided.aspx

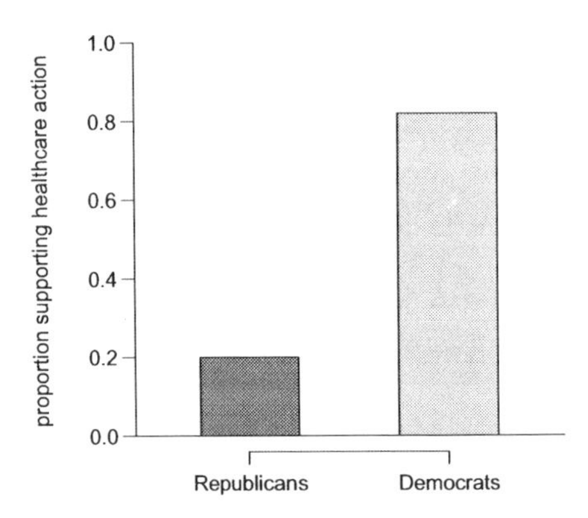

Figure 5.11: Support for Congressional action on Healthcare, by political party (Gallup poll, January 2010).

However, management must be convinced that the more expensive gears are worth the conversion before they approve the switch. If there is strong evidence that more than a 3% improvement in the percent of gears that pass inspection, management says they will switch suppliers, otherwise they will maintain the current supplier. Setup appropriate hypotheses for the test. Answer in the footnote[15].

● **Example 5.7** The quality control engineer from Exercise 5.12 collects a sample of gears, examining 1000 gears from each company and finds that 899 gears pass inspection from the current supplier and 958 pass inspection from the prospective supplier. Using this data, evaluate the hypothesis setup of Exercise 5.12 using a significance level of 5%.

First, conditions are checked. The sample is not necessarily random, so to proceed we must assume the gears are all independent; we will suppose this assumption is reasonable. The success-failure condition also holds for each sample. Thus, the difference in sample proportions, $0.958 - 0.899 = 0.059$, can be said to come from a nearly normal distribution.

The standard error can be found using Equation (5.4):

$$SE = \sqrt{\frac{0.958(1-0.958)}{1000} + \frac{0.899(1-0.899)}{1000}} = 0.0114$$

In this hypothesis test, the sample proportions were used. We will discuss this choice more in Section 5.4.3.

Next, we compute the test statistic and use it to find the p-value, which is depicted in Figure 5.12.

$$Z = \frac{\text{point estimate} - \text{null value}}{SE} = \frac{0.059 - 0.03}{0.0114} = 2.54$$

[15] H_0 : The higher quality gears will pass inspection no more than 3% more frequently than the standard quality gears. $p_{highQ} - p_{standard} = 0.03$. H_A : The higher quality gears will pass inspection more than 3% more often than the standard quality gears. $p_{highQ} - p_{standard} > 0.03$.

	cancer	noCancer
2,4-D	191	304
no 2,4-D	300	641

Table 5.13: Summary results for cancer in dogs and the use of 2,4-D by the dog's owner.

Using the normal model for this test statistic, we identify the right tail area as 0.006. Because this is a one-sided test, this single tail area is also the p-value, and we reject the null hypothesis because 0.006 is less than 0.05. That is, we have statistically significant evidence that the higher quality gears actually do pass inspection more than 3% as often as the currently used gears. Based on these results, management will approve the switch to the new supplier.

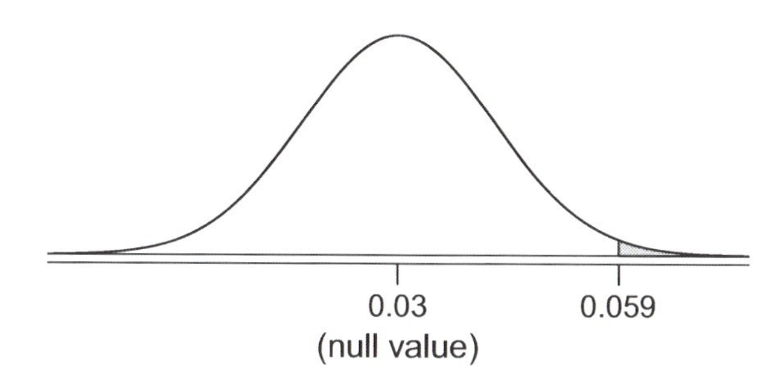

Figure 5.12: Distribution of the test statistic if the null hypothesis was true. The p-value is represented by the shaded area.

5.4.3 Hypothesis testing when $H_0 : p_1 = p_2$

Here we examine a special estimate of standard error when $H_0 : p_1 = p_2$ in the context of a new example.

We investigate whether there is an increased risk of cancer in dogs that are exposed to the herbicide 2,4-dichlorophenoxyacetic acid (2,4-D). A study in 1994 examined 491 dogs that had developed cancer and 945 dogs as a control group. Of these two groups, researchers identified which dogs had been exposed to 2,4-D in their owner's yard, and the results are shown in Table 5.13.

⊙ **Exercise 5.13** Is this study an experiment or an observational study? [16]

[16] The owners were not instructed to apply or not apply the herbicide, so this is an observational study. This question was especially tricky because one group was called the *control group*, which is a term usually seen in experiments.

⊙ **Exercise 5.14** Setup hypotheses to test whether 2,4-D and the occurrence of cancer in dogs is in anyway related. Use a one-sided test and compare across the cancer and no cancer groups. Comment and answer in the footnote[17].

● **Example 5.8** Are the conditions met to use the normal model and make inference on the results?

(1) It is unclear whether this is a random sample. However, if we believe the dogs in both the cancer and no cancer groups are representative of each respective population and that the dogs in the study do not interact in any way, then we may find it reasonable to assume independence between observations holds. (2) The success-failure condition holds for each sample.

Under the assumption of independence, we can use the normal model and make statements regarding the canine population based on the data.

In your hypothesis setup for Exercise 5.14, the null hypothesis is that the proportion of dogs with exposure to 2,4-D is the same in each group. The point estimate of the difference in sample proportions is $\hat{p}_c - \hat{p}_n = 0.067$. To identify the p-value for this test, we first check conditions (Example 5.8) and compute the standard error of the difference:

$$SE = \sqrt{\frac{p_c(1-p_c)}{n_c} + \frac{p_n(1-p_n)}{n_n}}$$

In a hypothesis test, the distribution of the test statistic is always examined as though the null hypothesis was true, i.e. in this case, $p_c = p_n$. The standard error formula should reflect this equality in the null hypothesis. We will use p to represent the common rate of dogs that are exposed to 2,4-D in the two groups:

$$SE = \sqrt{\frac{p(1-p)}{n_c} + \frac{p(1-p)}{n_n}}$$

We don't know the exposure rate, p, but we can obtain a good estimate of it by *pooling* the results of both samples:

$$\hat{p} = \frac{\text{\# of ``successes''}}{\text{\# of cases}} = \frac{191 + 304}{191 + 300 + 304 + 641} = 0.345$$

[17]Using the proportions within the `cancer` and `noCancer` groups rather than examining the rates for cancer in the `2,4-D` and `no 2,4-D` groups may seem odd; we might prefer to condition on the use of the herbicide, which is an explanatory variable in this case. However, the cancer rates in each group do not necessarily reflect the cancer rates in reality due to the way the data were collected. For this reason, computing cancer rates may greatly alarm dog owners.
H_0: the proportion of dogs with exposure to 2,4-D is the same in the `cancer` and `noCancer` groups ($p_c - p_n = 0$).
H_A: the dogs with cancer are more likely to have been exposed to 2,4-D than dogs without cancer ($p_c - p_n > 0$).

This is called the **pooled estimate** of the sample proportion, and we use it to compute the standard error when the null hypothesis is that $p_1 = p_2$ (e.g. $p_c = p_n$ or $p_c - p_n = 0$). We also typically use it to verify the success-failure condition.

Pooled estimate of a proportion
When the null hypothesis is $p_1 = p_2$, it is useful to find the pooled estimate of the shared proportion:

$$\hat{p} = \frac{\text{number of "successes"}}{\text{number of cases}} = \frac{\hat{p}_1 n_1 + \hat{p}_2 n_2}{n_1 + n_2}$$

Here $\hat{p}_1 n_1$ represents the number of successes in sample 1 since

$$\hat{p}_1 = \frac{\text{number of successes in sample 1}}{n_1}$$

Similarly, $\hat{p}_2 n_2$ represents the number of successes in sample 2.

TIP: Utilizing the pooled estimate of a proportion when $\mathbf{H_0 : p_1 = p_2}$
When the null hypothesis suggests the proportions are equal, we use the pooled proportion estimate ($\hat{p}$) to verify the success-failure condition and also to estimate the standard error:

$$SE = \sqrt{\frac{\hat{p}(1-\hat{p})}{n_c} + \frac{\hat{p}(1-\hat{p})}{n_n}} \tag{5.5}$$

⊙ **Exercise 5.15** Using Equation (5.5), verify the estimate for the standard error using $\hat{p} = 0.345$, $n_1 = 491$, and $n_2 = 945$ is $SE = 0.026$.

⊙ **Exercise 5.16** Using a significance level of 0.05, complete the hypothesis test using the standard error from Exercise 5.15. Be certain to draw a picture, compute the p-value, and state your conclusion in both statistical language and plain language. A short answer is provided in the footnote[18].

[18]Compute the test statistic:

$$Z = \frac{\text{point estimate} - \text{null value}}{SE} = \frac{0.067 - 0}{0.026} = 2.58$$

Looking this value up in the normal probability table: 0.9951. However this is the lower tail, and the upper tail represents the p-value: $1 - 0.9951 = 0.0049$. We reject the null hypothesis and conclude that dogs getting cancer and owners using 2,4-D are associated.

5.5 When to retreat

The conditions described in each section are to help ensure the point estimates are nearly normal and unbiased. When the conditions are not met, these methods cannot be safely applied or drawing conclusions from the data may be treacherous. The conditions for each test typically come in two forms.

- The individual observations must be independent. A random sample from less than 10% of the population ensures the observations are independent. In experiments, other considerations must be made to ensure observations are independent. If independence fails, then advanced techniques must be used, and in some such cases, inference regarding the target population may not be possible.
- Other conditions focus on sample size and skew. If the sample size is too small or the skew too strong, then the normal model may again fail.

For analyzing smaller samples where the normal model may not apply, we refer to Chapter 6.

While we emphasize special caution here, verification of conditions for statistical tools is always necessary. Whenever conditions are not satisfied for a statistical tool, there are three options. The first is to learn new methods that are appropriate for the data. The second route is to hire a statistician[19]. The third route is to ignore the failure of conditions. This last option effectively invalidates any analysis and may discredit novel and interesting findings. For the respected scientist, this last option is not viable.

Finally, we caution that there may be no inference tools helpful when considering data that includes unknown biases, such as convenience samples. For this reason, there are books, courses, and researchers devoted to the techniques of sample and experimental design. We consider some basic methods in Chapter 10. [Authors' note: this chapter number may change.]

[19] If you work at a university, then there may be campus consulting services to assist you. Alternatively, there are many private consulting firms that are also available for hire.

5.6 Problem set

5.6.1 Paired data

5.6.2 Difference of two means

5.1 In 1964 the Surgeon General released their first report linking smoking to various health issues, including cancer. Research done by Sterling Cooper, a prestigious ad agency in New York City, showed that before the report was released, in a random sample of 80 smokers, the average number of cigarettes smoked per day was 13.5 with a standard deviation of 3.2. A year after the report was released, in a random sample of 85 smokers, the average number of cigarettes smoked per day was 12.6 with a standard deviation of 2.9. Is there evidence to suggest that the average number of cigarettes smoked per day has decreased after the Surgeon General's report?

	Before	After
n	80	85
$\bar{x}$	13.5	12.6
s	3.2	2.9

(a) Write the hypotheses in symbols and in words.

(b) Are the assumptions/conditions for inference satisfied?

(c) Calculate the test statistic and find the p-value.

(d) What do you conclude? Interpret your conclusion in context.

(e) Does this imply that the Surgeon General's report was the cause of this decrease? Explain.

(f) What type of error might we have committed? Explain.

5.2 Based on the data given in Exercise 1, construct a 90% confidence interval for the difference between the average number of cigarettes smoked per day before and after the Surgeon General's report was released. Interpret this interval in context. Also comment on if the confidence interval agree with the conclusion of the hypothesis test from Exercise 1.

5.3 An article published in the International Journal of Obesity [10] examined the accuracy of BMI in diagnosing obesity. Data from 13,601 subjects aged 20-79.9 from the Third National Health and Nutrition Examination Survey were studied. Two of the variables of interest were body fat percentage (BF) and lean mass. Below is an excerpt of some of results of this study. Note that BMI, BF and lean mass are given as *mean* $\pm$ *standard error*. Assume that all assumptions and conditions for inference are satisfied.

Gender	n	BF (%)	Lean mass (kg)
Men	6580	23.9 ± 0.07	61.8 ± 0.12
Women	7021	35.0 ± 0.09	44.0 ± 0.08

(a) One definition of obesity in women is greater than 35% body fat and in men greater than 25%. The cutoff for women is higher since women store extra fat in their bodies to be used during childbearing. This implies that women on average should have a higher body fat percentage than men. Test this hypothesis using $\alpha = 0.01$.

(b) Lean mass and fat mass are what makes up a person's total weight. In part (a) we showed that women on average have higher body fat percentage. The data also show that women have lower mean mass. Is the only reason for this higher body fat percentage, or can you think of other reasons why women on average have lower lean mass?

5.6.3 Single population proportion

5.4 We are interested in estimating the proportion of graduates at a mid-sized university who found a job within one year of completing their undergraduate degree. We conduct a survey and find out that out of the 400 randomly sampled graduates 348 found jobs within within one year of completing their undergraduate degree.

(a) Define the sample statistic and the population parameter of interest. What is the value of the sample statistic?

(b) Are assumptions and conditions for inference satisfied?

(c) Construct a 95% confidence interval for the proportion of graduates who who found a job within one year of completing their undergraduate degree at this university.

(d) Explain what this interval means in the context of this question.

(e) What does "95% confidence" mean?

(f) If we increased the confidence level what would happen to the width of the interval, i.e. how would the precision of the interval change. (*Hint: You do not need to calculate the interval to answer this question.*)

(g) If we increased the sample size what would happen to the width of the interval, i.e. how would the precision of the interval change. (*Hint: You do not need to calculate the interval to answer this question.*)

5.5 A U.S. university that recently implemented a new study abroad program conducted a campus wide survey to find out what percent of the students have travelled abroad. The survey results showed that out of the 100 randomly sampled students at this university, 42 have travelled outside the U.S.

(a) Define the sample statistic and the population parameter of interest. What is the value of the sample statistic?

(b) Are assumptions and conditions for inference satisfied?

(c) Construct a 90% confidence interval for the proportion of students at this university who have travelled abroad.

(d) Interpret this interval in context.

(e) What does "90% confidence" mean?

5.6 Exercise 5 provides the result of a campus wide survey which showed that 42 out of the 100 randomly sampled students have travelled abroad. Prior to the implementation of the study abroad program a study from was conducted at the same university which had shown that 35% of the students had travelled abroad. Is there significant evidence to suggest that the proportion of students at this university who have travelled abroad has increased after the implementation of the study abroad program?

(a) Conduct a hypothesis test to support your answer.

(b) Interpret the p-value in context.

(c) Does the conlcusion of your hypothesis test agree with the confidence interval constructed in Exercise 5?

5.7 A college review magazine states that in many business schools there is a certain stigma that marketing is an easy major and that therefore majority of students majoring in marketing also major in finance, economics or accounting to be able to show employers that their quantitative skills are also strong. In order to test this claim an education researcher collects a random sample of 80 undergraduate students majoring in marketing at various business schools, and finds that 50 of them have a double major. Is there evidence to support the magazine's claim that majority of marketing students have a double major?

(a) Write the hypotheses in symbols and in words.

(b) Are assumptions and conditions for inference satisfied?

(c) Calculate the test statistic.

(d) Find and interpret the p-value in context.

(e) Based on the hypothesis test, do the data suggest that majority of marketing students have a double major? Explain.

5.8 Based on the information given in Exercise 7, Construct a 90% confidence interval for the proportion of marketing students who have a double major. Interpret this interval in context. Also comment on if the confidence interval agree with the conclusion of the hypothesis test from Exercise 7.

5.9 Some people claim that they can tell the difference between a diet soda and a regular soda in the first sip. A researcher wanting to test this claim randomly sampled 80 such people. He then filled 80 plain white cups with soda, half diet and half regular through random assignment, and asked each person to take one sip from their cup and identify the soda as "diet" or "regular". 53 participants correctly identified the soda. Does this suggest that these people are able to detect the difference between diet and regular soda, in other words, are the results significantly better than just random guessing?

(a) Write the hypotheses in symbols and in words.

(b) Are the assumptions and conditions for inference satisfied?

(c) Calculate the test statistic.

(d) Find and interpret the p-value in context.

(e) Based on the hypothesis test, do the data suggest the rate of correctly identifying a soda for these people is better than just by random guessing?

5.10 According to the Centers for Disease Control and Prevention, 30% of Americans are habitually getting less than six hours of sleep a night - far less than the recommended seven to nine hours. New York is known as "the city that never sleeps". In a random sample of 300 New Yorkers it was found that 105 of them get less than six hours of sleep a night. Is there significant evidence to suggest that the rate of sleep deprivation for New Yorkers is higher than the rate of sleep deprivation in the population at large?

(a) Write the hypotheses in symbols and in words.

(b) Are the assumptions and conditions for inference satisfied?

(c) Calculate the test statistic.

(d) Find and interpret the p-value in context.

(e) Based on the hypothesis test, do the data suggest the rate of sleep deprivation for New Yorkers is higher than the rate of sleep deprivation in the population at large? Explain.

5.11 We are interested in estimating the proportion of students at a university who smoke. Out of a random sample of 200 students from this university, 40 students smoke.

(a) Are assumptions and conditions for inference satisfied?

(b) Construct a 95% confidence interval for the proportion of students at this university who smoke, and interpret this interval in context.

(c) If we wanted the margin of error to be no larger than 4% for a 95% confidence interval for the proportion of students who smoke, how big a sample would we need?

(d) Construct a 99% confidence interval for the proportion of students at this university who **do not** smoke, and interpret this interval in context.

5.12 A survey conducted five years ago by the health center at the university mentioned in Exercise 11 showed that 18% of the students smoked.

(a) Is there evidence to suggest that the percentage of students who smoke has changed over the last five years?

(b) What type of error might we have committed above?

5.13 It is believed that large doses of acetaminophen (the active ingredient in over the counter pain relievers like Tylenol) may cause damage to the liver. A researcher wants to conduct a study to estimate the proportion of acetaminophen users who have liver damage. For participating in this study she will pay each subject $20.

(a) If she wants to limit the margin of error of her 98% confidence interval to 2%. What is the minimum amount of money she needs to set aside to pay her subjects?

(b) The amount you calculated in part (a) is way over her budget so she decides to use fewer subjects. How will this affect the width of her confidence interval, i.e. how will the precision of her confidence interval change? (*Hint: You do not need to calculate the interval to answer this question.*)

5.14 Michael Scott, a regional manager at Dunder Mifflin Paper Company, claims that over 45% of businesses have Dunder Mifflin as their sole paper provider. To test his claim we randomly sampled 180 businesses and found that 36% of these businesses had Dunder Mifflin as their sole paper provider. Is there significant evidence to support Michael Scott's claim that over 45% of businesses have Dunder Mifflin as their sole paper provider?

(a) Write the hypotheses in symbols and in words.

(b) Are the assumptions and conditions for inference satisfied?

(c) Calculate the test statistic.

(d) Find and interpret the p-value in context.

(e) What do you conclude? Interpret your conclusion in context.

5.15 Statistics show that traditionally about 65% of Kansas residents go out of state for college. This year researchers wanting to see if this number has increased randomly sampled 1,500 high-school graduates and found that 1,005 went out of state for college.

(a) Write the hypotheses in symbols and in words.

(b) Are the assumptions and conditions for inference satisfied?

(c) Calculate the test statistic.

(d) Find and interpret the p-value in context.

(e) Based on the hypothesis test is there significant evidence to suggest that the percentage of Kansas residents who go out of state for college has increased? Explain.

5.16 A Washington Post article [11] reports that "a new Washington Post-ABC News poll shows that support for a government-run health-care plan to compete with private insurers has rebounded from its summertime lows and wins clear majority support from the public." More specifically the article says "seven in 10 Democrats back the plan, while almost nine in 10 Republicans oppose it. Independents divide 52 percent against, 42 percent in favor of the legislation." There were were 819 democrats, 566 republicans and 783 independents surveyed.

A political pundit on TV claims that democrats should not count on the independent votes for the public option plan since more than half of the independents oppose it. Is there significant evidence to support the pundit's claim?

(a) Write the hypotheses in symbols and in words.

(b) Are the assumptions and conditions for inference satisfied?

(c) Calculate the test statistic.

(d) Find and interpret the p-value in context.

(e) Based on the hypothesis test is there significant evidence to suggest that more than half of the independents oppose the public option plan? Explain. no, fail to reject null

(f) Would you expect a confidence interval for the proportion of independents who oppose the public option plan to include 0.5? Explain.

95% C.I. : point est. ± z* SE = 0.52 ± 1.96 · 0.01787 = [0.485, 0.555]

5.17 Exercise 16 presents the results of a political poll on support for the public option plan.

(a) Calculate a 90% confidence interval for the proportion of independents who oppose the public option plan.

(b) Interpret the confidence interval in context.

(c) Does your interval agree with the result of your hypothesis test from Exercise 16? Explain.

(d) If we wanted to limit the margin of error of a 90% confidence interval to 1%, how many independents would we need to survey?

5.6.4 Difference of two proportions

5.18 Exercise 15 mentions that out of a randomly sample of 1,500 high-school graduates from Kansas 1,005 went out of state for college. A similar survey conducted in California shows that from a random sample of 1,250 California high-school graduates 600 went out of state for college.

(a) Calculate a 90% confidence interval for the difference between the population proportions of Kansas and California residents who go out of state for college. Do not forget to check that assumptions and conditions for inference are met.

(b) Interpret the confidence interval in context.

(c) Based on your confidence interval, is there significant evidence to suggest that a higher proportion of Kansas residents go out of state for college? Explain.

(d) Could we use this interval as a proof that the college system is better in California and hence a higher proportion of California residents stay in the state for college? Explain.

5.19 Exercise 18 presents the results of two surveys conducted in Kansas and California. Conduct a hypothesis test to determine if there is a significant evidence to suggest that

(a) a higher proportion of Kansas residents go out of state for college,

(b) proportion of Kansas and California residents who go out of state for college are different.

5.20 Based on the information given in Exercise 16, is there significant evidence to suggest that a higher proportion of democrats than independents support the public option plan?

(a) Write the hypotheses in words and in symbols.

(b) Calculate the test statistic.

(c) Find the p-value and interpret it in context.

(d) What do you conclude?

(e) What type of error might we have committed?

(f) Would you expect a confidence interval for the difference between the two proportions to include 0? Explain your reasoning. If you answered no, would you expect the confidence interval for $(p_D - p_I)$ to be positive or negative?

(g) Calculate a 90% confidence interval for the difference between $(p_D - p_I)$.

(h) Interpret the confidence interval in context.

(i) What does "90% confident" mean?

(j) Does this imply that being a Democrat causes someone to support the public option plan? Explain.

5.21 According to a report on sleep deprivation by the Centers for Disease Control and Prevention [12] the proportion of California residents who reported insufficient rest or sleep during the on each of the preceding 30 days is 8.0% while this proportion is 8.8% for Oregon residents. A random sample of 11,545 California and 4,691 Oregon residents were surveyed. We are interested in finding out if there evidence to suggest that the rate of sleep deprivation is different for the two states.

(a) What kind of study is this?

(b) What are the hypotheses?

(c) Are the assumptions and conditions for inference satisfied?

(d) Calculate the test statistic.

(e) Find and interpret the p-value in context.

(f) What do you conclude? Interpret your conclusion in context.

(g) Does this imply that the rate of sleep deprivation is equal in the two states? Explain.

(h) What type of error might we have committed?

(i) Would you expect a confidence interval for the difference between the two proportions to include 0? Explain your reasoning.

5.22 Using the data provided in Exercise 21, construct a 95% confidence interval for the difference between the population proportions. Interpret the confidence interval in context and comment on if the confidence interval agree with the conclusion of the hypothesis test from Exercise 21.

5.23 Do the result of your hypothesis test from Exercise 21 agree with the interpretation of the confidence interval from Exercise 22? If not, why might that be the case?

5.6.5 When to retreat

5.24 The Stanford University Heart Transplant Study was conducted to determine whether an experimental heart transplant program increased lifespan [6]. Each patient entering the program was designated officially a heart transplant candidate, meaning that he was gravely ill and would most likely benefit from a new heart. Some patients got a transplant (treatment group) and some did not (control group). The table below displays how many patients survived and died in each group.

	control	treatment
alive	4	24
dead	30	45

Construct a 95% confidence interval for the difference between the proportion of patients who survived in the treatment and control groups. Do not forget to check that assumptions and conditions for inference are satisfied.

Chapter 6

Small sample inference

Large samples are sometimes unavailable, so it is useful to study methods that apply to small samples. Moving from large samples to small samples creates a number of problems that prevent us from applying the normal model directly, though we can still use the ideas from Chapter 4. We generalize the approach is as follows:

- Determine what test statistic is useful.
- Identify the distribution of the test statistic under the condition the null hypothesis was true.
- Apply the methods of Chapter 4 under the new distribution.

This is the same approach we used in Chapter 5. It just so happened that the normal model worked very well for all the point estimates we studied when the sample size was large.

[Authors' note: This chapter will include simulation, randomization, and other more precise techniques in the 2011 release.]

6.1 Small sample inference for the mean

We applied a normal model to the sample mean in Chapter 4 when the (1) observations were independent, (2) sample size was at least 50, and (3) the data were not strongly skewed. The findings in Section 4.4 also suggested we could relax condition (3) when we considered ever larger samples.

In this section, we examine the distribution of the sample mean for any sample size. To this end, we must strengthen the condition about the distribution of the data. Specifically, our data must meet two criteria:

(1) The observations are independent.

(2) The data come from a nearly normal distribution.

If we are not confident that the data are nearly normal, then we cannot apply the methods of this section. We can relax this condition as the sample size becomes larger.

It is useful to revisit our reasons for requiring a large sample for the normal model. We first wanted to ensure the sample mean was nearly normal. Secondly, when the sample size was small, the standard error estimate may not reliable. A sample size of 50 or more helps ensure this error estimate was adequate. We examine these two issues separately, leading us to a new distribution that will be useful for small sample inference about means.

6.1.1 The normality condition

If the individual observations are independent and come from a nearly normal distribution, a special case of the Central Limit Theorem ensures the distribution of the sample means is nearly normal.

> **Central Limit Theorem for normal data**
> The sample mean is nearly normal when the observations are independent and come from a nearly normal distribution. This is true for any sample size.

This special case of the Central Limit Theorem resolves concerns about the normality of the mean when the sample size is small and the data is also nearly normal. However, we encounter a new problem: it is inherently difficult to verify normality in small data sets.

> **Caution: Checking the normality condition**
> We should exercise caution when verifying the normality condition for small samples. It is important to not only examine the data but also think about where the data come from. Always ask: Is it reasonable to assume this type of data is nearly normal?

6.1.2 Introducing the t distribution

We will address the uncertainty of the standard error estimate by using a new distribution: the t distribution. A t distribution, shown as a solid line in Figure 6.1,

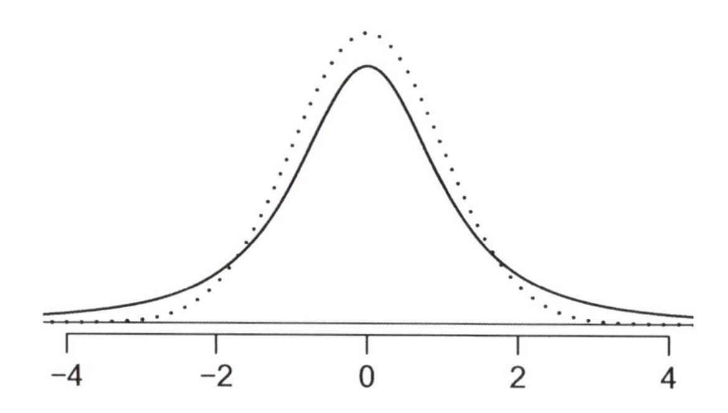

Figure 6.1: Comparison of a t distribution (solid line) and a normal distribution (dotted line).

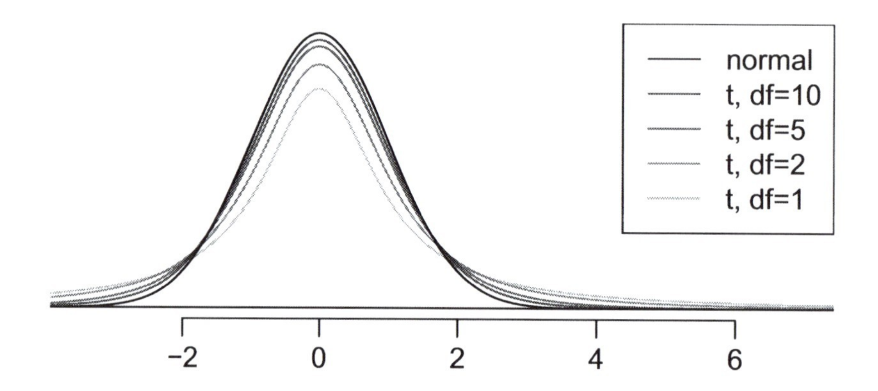

Figure 6.2: The larger the degrees of freedom, the more closely the t distribution resembles the standard normal model.

has a bell shape. However, its tails are thicker than the normal model's. This means observations are more likely to fall beyond two standard deviations from the mean than under the normal distribution[1]. These extra thick tails are exactly the correction we need to resolve our problem with estimating the standard error.

The t distribution, always centered at zero, has a single parameter: degrees of freedom. The **degrees of freedom (df)** describe the exact bell shape of the t distribution. Several t distributions are shown in Figure 6.2. When there are more degrees of freedom, the t distribution looks very much like the standard normal distribution.

[1]The standard deviation of the t distribution is actually a little more than 1. However, it is useful to always think of the t distribution as having a standard deviation of 1 in all of our applications.

Degrees of freedom
The degrees of freedom describe the shape of the t distribution. The larger the degrees of freedom, the more closely the distribution approximates the normal model.

When the degrees of freedom is about 50 or more, the t distribution is nearly indistinguishable from the normal distribution. In Section 6.1.4, we relate degrees of freedom to sample size.

6.1.3 Working with the t distribution

We will find it very useful to become familiar with the t distribution because it plays a very similar role to the normal distribution during inference. It will be useful to have a t table that can be used in place of the normal probability table. This **t table** is partially shown in Table 6.3. A larger table is presented in Appendix A.2 on page 250.

one tail	0.100	0.050	0.025	0.010	0.005
two tails	0.200	0.100	0.050	0.020	0.010
df 1	3.08	6.31	12.71	31.82	63.66
2	1.89	2.92	4.30	6.96	9.92
3	1.64	2.35	3.18	4.54	5.84
⋮	⋮	⋮	⋮	⋮	
17	1.33	1.74	2.11	2.57	2.90
18	*1.33*	*1.73*	*2.10*	*2.55*	*2.88*
19	1.33	1.73	2.09	2.54	2.86
20	1.33	1.72	2.09	2.53	2.85
⋮	⋮	⋮	⋮	⋮	
400	1.28	1.65	1.97	2.34	2.59
500	1.28	1.65	1.96	2.33	2.59
∞	1.28	1.64	1.96	2.33	2.58

Table 6.3: An abbreviated look at the t table. Each row represents a different t distribution. The columns describe the tail areas at each standard deviation. The row with df = 18 has been *highlighted*.

Each row in the t table represents a t distribution with different degrees of freedom. The columns represent values corresponding to tail probabilities. For instance, if we know we are working with the t distribution with df = 18, we can examine row 18, which is highlighted in Table 6.3. If we want the value in this row that identifies the cutoff for an upper tail of 10%, we can look in the column where *one tail* is 0.100. This cutoff is 1.33. If we had wanted the cutoff for the lower

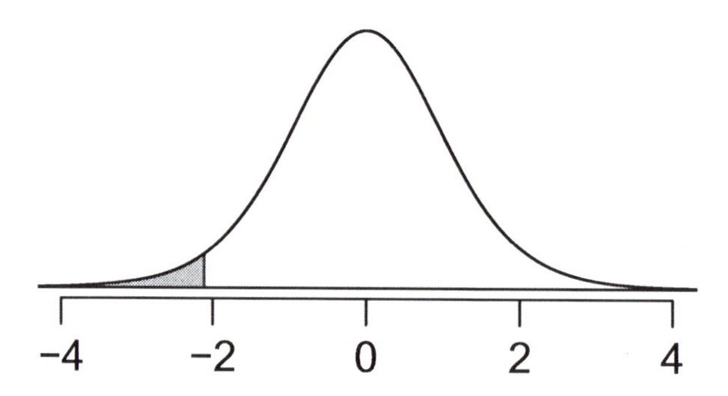

Figure 6.4: The t distribution with 18 degrees of freedom. The area below -2.10 has been shaded.

10%, we would use -1.33; just like the normal distribution, all t distributions are symmetric.

- **Example 6.1** What proportion of the t distribution with 18 degrees of freedom falls below -2.10?

  Just like a normal probability problem, we first draw the picture in Figure 6.4. We seek the area below -2.10, which is shaded in the picture. To find this area, we first identify the appropriate row: $df = 18$. Then we identify the column containing the absolute value of -2.10: the third column. Because we are looking for just one tail, we examine the top line of the table, which shows that a one tail area for a value in the third row corresponds to 0.025. About 2.5% of the distribution falls below -2.10. In the next example we encounter a case where the exact t value is not listed in the table.

- **Example 6.2** A t distribution with 20 degrees of freedom is shown in the left panel of Figure 6.5. Estimate the proportion of the distribution falling above 1.65.

  We identify the row in the t table using the degrees of freedom: $df = 20$. Then we look for 1.65; it is not listed. It falls between the first and second columns. Since these values bound 1.65, their tail areas will bound the tail area corresponding to 1.65. We identify the one tail area of the first and second columns, 0.050 and 0.10, and we conclude that between 5% and 10% of the distribution is more than 1.65 standard deviations above the mean. If we like, we can identify the precise area using a computer with statistical software: 0.0573.

- **Example 6.3** A t distribution with 2 degrees of freedom is shown in the right panel of Figure 6.5. Estimate the proportion of the distribution falling more than 3 units from the mean (above or below).

  As before, first identify the appropriate row: $df = 2$. Next, find the columns that capture 3; because $2.92 < 3 < 4.30$, we use the second and third columns. Finally,

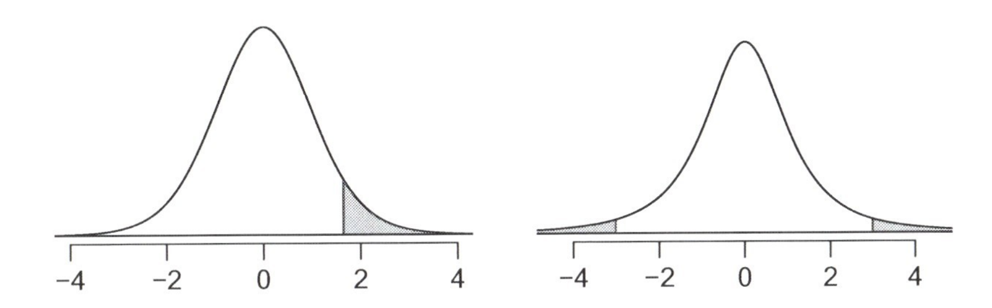

Figure 6.5: Left: The t distribution with 20 degrees of freedom, with the area above 1.65 shaded. Right: The t distribution with 2 degrees of freedom, and the area further than 3 units from 0 has been shaded.

we find bounds for the tail areas by looking at the two tail values: 0.05 and 0.10. We use the two tail values because we are looking for two (symmetric) tails.

6.1.4 The t distribution as a solution to the standard error problem

When estimating the mean and standard error from a small sample, the t distribution is a more accurate tool than the normal model.

> **TIP: When to use the t distribution**
> When observations are independent and nearly normal, we can use the t distribution for inference of the sample mean and estimated standard error in the same way we used the normal model in Chapters 4 and 5.

We use the t distribution instead of the normal model because we have extra uncertainty in the estimate of the standard error. To proceed with the t distribution for inference about a single mean, we must check two conditions.

- Independence of observations: We verify this condition exactly as before. We either collect a simple random sample from less than 10% of the population or, if it was an experiment or random process, carefully ensure to the best of our abilities that the observations were independent.
- Observations come from a nearly normal distribution: This second condition is more difficult to verify since we are usually working with small data sets. Instead we often (i) take a look at a plot of the data for obvious deviations from the normal model and (ii) consider whether any previous experiences alert us that the data may not be nearly normal.

When examining a sample mean and estimated standard error from a sample of n independent and nearly normal observations, we will use a t distribution with $n-1$ degrees of freedom (df). For example, if the sample size was 19, then we would use the t distribution with $df = 19 - 1 = 18$ degrees of freedom and proceed exactly as we did in Chapter 4 *but now we use the t table.*

Figure 6.6: A Risso's dolphin. Photo by Mike Baird. Image is under Creative Commons Attribution 2.0 Generic.

	n	$\bar{x}$	s	minimum	maximum
sample summaries	19	4.4	2.3	1.7	9.2

Table 6.7: Summary of mercury content in the muscle of 19 Risso's dolphins from the Taiji area. Measurements are in μg/wet g (micrograms of mercury per wet gram of muscle).

We can relax the normality condition for the observations when the sample size becomes large. For instance, a slightly skewed data set might be acceptable if there were at least 15 observations. For a strongly skewed data set, we might require 30 or 40 observations. For an extremely skewed data set, perhaps 100 or more.

6.1.5 One sample confidence intervals with small n

Dolphins are at the top of the oceanic food chain, which causes dangerous substances such as mercury to concentrate in their organs and muscles. This is an important problem for both dolphins and other animals, like humans, who occasionally eat them. For instance, this is particularly relevant in Japan where school meals have included dolphin at times.

Here we identify a confidence interval for the average mercury content in dolphin muscle using a sample of 19 Risso's dolphins from the Taiji area in Japan[2]. The data are summarized in Table 6.7. The minimum and maximum observed values can be used to evaluate whether or not there are any extreme outliers or obvious skew.

⊙ **Exercise 6.1** Are the independence and normality conditions satisfied for this data

[2]Taiji was featured in the movie *The Cove*, and it is a significant source of dolphin and whale meat in Japan. Thousands of dolphins pass through the Taiji area annually, and we will assume these 19 dolphins represent a random sample from those dolphins. Data reference: Endo, t and Haraguchi, K (2009). *High mercury levels in hair samples from residents of Taiji, a Japanese whaling town.* Marine Pollution Bulletin. In press.

set? [3]

In the normal model, we used z^* and the standard error to determine the width of a confidence interval. When we have a small sample, we try the t distribution instead of the normal model:

$$\bar{x} \pm t^*_{df} SE$$

The sample mean and estimated standard error are computed just as before ($\bar{x} = 4.4$ and $SE = s/\sqrt{n} = 0.528$), while the value t^*_{df} is a change from our previous formula. Here t^*_{df} corresponds to the appropriate cutoff from the t distribution with df degrees of freedom, which is identified below.

> **Degrees of freedom for a single sample**
> If our sample has n observations and we are examining a single mean, then we use the t distribution with $df = n - 1$ degrees of freedom.

The rule states that we should use the t distribution with $df = 19 - 1 = 18$ degrees of freedom. When using a 95% confidence interval, this corresponds to 2.5% in the lower tail and 2.5% in the upper tail of the normal model. It is the same with the t distribution. We can look in the abbreviated t table on page 203 where each tail has 2.5% (both tails total to 5%), which is the third column. Then we identify the row with 18 degrees of freedom to obtain $t^*_{18} = 2.10$. Generally the t^*_{df} is slightly larger than what we would expect under the normal model with z^*.

Finally, we can substitute all our values in to create the 95% confidence interval for the average mercury content in muscles from Risso's dolphins that pass through the Taiji area:

$$\bar{x} \pm t^*_{18} SE \quad \to \quad 4.4 \pm 2.10 * 0.528 \quad \to \quad (3.87, 4.93)$$

We are 95% confident the average mercury content of muscles in Risso's dolphins is between 3.87 and 4.93 μg/wet gram. The US safety limit is 0.5 μg per wet gram[4].

> **Finding a t confidence interval for the mean**
> Based on a sample of n independent and nearly normal observations, a confidence interval for the population mean is
>
> $$\bar{x} \pm t^*_{df} SE$$
>
> where $\bar{x}$ is the sample mean, t^*_{df} corresponds to the confidence level and degrees of freedom, and SE is the standard error as estimated by the sample.

[3] The observations are a random sample and consist of less than 10% of the population, therefore independence is reasonable. The summary statistics in Table 6.7 do not suggest any strong skew or outliers, which is encouraging. Based on this evidence – and that we don't have any clear reasons to believe the data are not roughly normal – the normality assumption is reasonable.

[4] http://www.ban.org/ban-hg-wg/Mercury.ToxicTimeBomb.Final.PDF

⊙ **Exercise 6.2** The FDA's webpage provides some data on mercury content of fish[5]. Based on a sample of 15 croaker white fish (Pacific), a sample mean and standard deviation were computed as 0.287 and 0.069 ppm (parts per million), respectively. The 15 observations ranged from 0.18 to 0.41 ppm. We will assume these observations are independent. Based on the summary statistics of the data, do you have any objections to the normality condition of the individual observations? [6]

● **Example 6.4** Estimate the standard error of $\bar{x} = 0.287$ ppm from the statistics in Exercise 6.2. If we are to use the t distribution to create a 90% confidence interval for the actual mean of the mercury content, identify the degrees of freedom we should use and also find t^*_{df}.

$SE = \frac{0.069}{\sqrt{15}} = 0.0178$ and $df = n - 1 = 14$. Looking in the column where two tails is 0.100 (since we want a 90% confidence interval) and row $df = 14$, we identify $t^*_{14} = 1.76$.

⊙ **Exercise 6.3** Based on the results of Exercise 6.2 and Example 6.4, compute a 90% confidence interval for the average mercury content of croaker white fish (Pacific). [7]

6.1.6 One sample t tests with small n

An SAT preparation company claims that its students' scores improve by over 100 points on average after their course. A consumer group would like to evaluate this claim, and they collect data on a random sample of 30 students who took the class. Each of these students took the SAT before and after taking the company's course, and we would like to examine the differences in these scores to evaluate the company's claim[8]. The distribution of the difference in scores, shown in Figure 6.8, has mean 135.9 and standard deviation 82.2. Does this data provide convincing evidence to back up the company's claim?

⊙ **Exercise 6.4** Setup hypotheses to evaluate the company's claim. Use μ to represent the true average difference in student scores. [9]

---

[5] http://www.fda.gov/Food/FoodSafety/Product-SpecificInformation/Seafood/FoodbornePathogensContaminants/Methylmercury/ucm115644.htm

[6] There are no extreme outliers; all observations are within 2 standard deviations of the mean. If there is skew, it is not strong. There are no red flags for the normal model based on this (limited) information, and we do not have reason to believe the mercury content is not nearly normal in this type of fish.

[7] Use $\bar{x} \pm t^*_{14}SE$: $0.287 \pm 1.76 * 0.0178$. This corresponds to $(0.256, 0.318)$. We are 90% confident that the average mercury content of croaker white fish (Pacific) is between 0.256 and 0.318 ppm.

[8] This was originally paired data and so we look at the differences. (See Section 5.1.)

[9] This is a one-sided test. H_0: student scores do not improve by more than 100 after taking the company's course. $\mu \leq 100$ (or simply $\mu = 100$). H_A: students scores improve by more than 100 points on average after taking the company's course. $\mu > 100$.

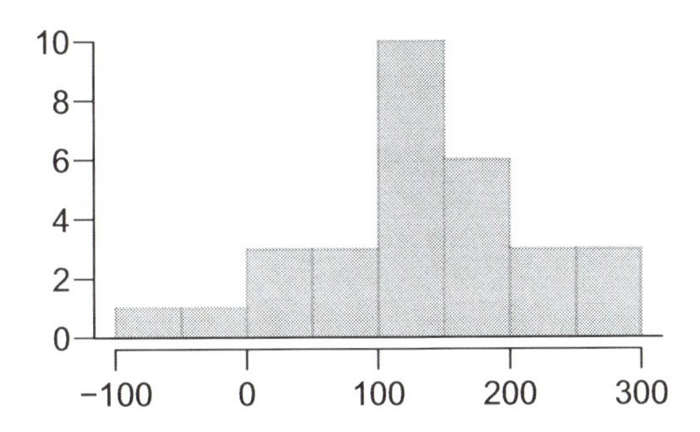

Figure 6.8: Sample distribution of improvements in SAT scores after taking the SAT course.

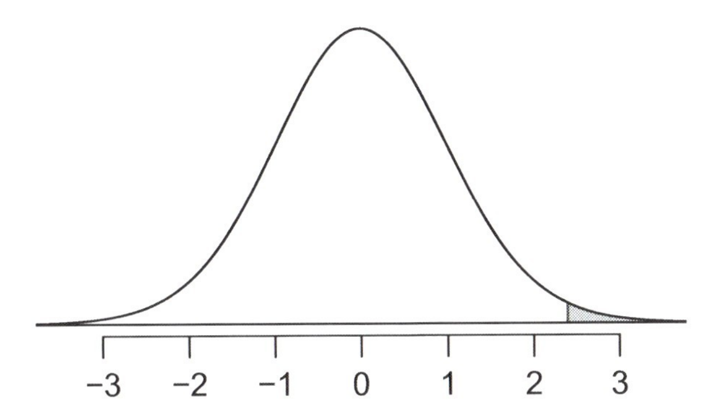

Figure 6.9: The t distribution with 29 degrees of freedom.

⊙ **Exercise 6.5** Are the conditions to use the t distribution method satisfied? [10]

Just as we did for the normal case, we standardize the sample mean using the Z score to identify the test statistic. However, we will write T instead of Z, because we have a small sample and are basing our inference on the t distribution:

$$T = \frac{\bar{x} - \text{null value}}{SE} = \frac{135.9 - 100}{82.2/\sqrt{30}} = 2.39$$

If the null hypothesis was true, the test statistic T would follow a t distribution with $df = n - 1 = 29$ degrees of freedom. We can draw a picture of this distribution and mark the observed T, as in Figure 6.9. The shaded right tail represents the p-value: the probability of observing such strong evidence favoring the SAT company's claim if the students only actually improved 100 points on average.

[10] This is a random sample from less than 10% of the company's students (assuming they have more than 300 former students), so the independence condition is reasonable. The normality condition also seems reasonable based on Figure 6.8. We can use the t distribution method.

⊙ **Exercise 6.6** Use the t table in Appendix A.2 on page 250 to identify the p-value. What do you conclude? [11]

⊙ **Exercise 6.7** Because we rejected the null hypothesis, does this mean that taking the company's class improves student scores by more than 100 points on average? [12]

6.2 The t distribution for the difference of two means

It is useful to be able to compare two means. For instance, a teacher might like to test the notion that two versions of an exam were equally difficult. She could do so by randomly assigning each version to students. If she found that the average scores on the exams were so different that we cannot write it off as chance, then she may want to award extra points to students who took the more difficult exam.

In a medical context, we might investigate whether embryonic stem cells (ESCs) can improve heart pumping capacity in individuals who have suffered a heart attack. We could look for evidence of greater heart health in the ESC group against the control group.

The ability to make conclusions about a difference in two means, $\mu_1 - \mu_2$, is often useful. If the sample sizes are small and the data is nearly normal, the t distribution can be applied to the sample difference in means, $\bar{x}_1 - \bar{x}_2$, to make inference about the difference in population means.

6.2.1 A distribution for the difference of two sample means

In the example of two exam versions, the teacher would like to evaluate whether there is convincing evidence that the difference in average scores is not due to chance.

It will be useful to extend the t distribution method from Section 6.1 to apply to a new point estimate:

$$\bar{x}_1 - \bar{x}_2$$

Just as we did in Section 5.2, we verify conditions for each sample separately and then verify that the samples are also independent. For example, if the teacher believes students in her class are independent, the exam scores are nearly normal, and the students taking each version of the exam were independent, then we can use the t distribution for the point estimate, $\bar{x}_1 - \bar{x}_2$.

[11] We use the row with 29 degrees of freedom. The value $T = 2.39$ falls between the third and fourth columns. Because we are looking for a single tail, this corresponds to a p-value between 0.01 and 0.025. The p-value is guaranteed to be less than 0.05 (the default significance level), so we reject the null hypothesis. The data provides convincing evidence to support the company's claim that student scores improve by more than 100 points following the class.

[12] This is an observational study, so we cannot make this causal conclusion. For instance, maybe SAT test takers tend to improve their score over time even if they don't take a special SAT class, or perhaps only the most motivated students take such SAT courses.

The formula for the standard error of $\bar{x}_1 - \bar{x}_2$, introduced in Section 5.2, remains useful for small samples:

$$SE_{\bar{x}_1 - \bar{x}_2} = \sqrt{SE^2_{\bar{x}_1} + SE^2_{\bar{x}_2}} = \sqrt{\frac{s_1^2}{n_1} + \frac{s_2^2}{n_2}} \qquad (6.1)$$

Because we will use the t distribution, we will need to identify the appropriate degrees of freedom. This can be done by using computer software. An alternative technique to a computer is to use the smaller of $n_1 - 1$ and $n_2 - 1$, which is the method we will use in the examples and exercises[13].

> **Using the t distribution for a difference in means**
> The t distribution can be used for the (standardized) difference of two means if (1) each sample meets the conditions for the t distribution and (2) the samples are independent. We estimate the standard error of the difference of two means using Equation (6.1).

6.2.2 Two sample t test

Summary statistics for each exam version are shown in Table 6.10. The teacher would like to evaluate whether this difference is so large that it provides convincing evidence that Version B was more difficult (on average) than Version A.

Version	n	$\bar{x}$	s	min	max
A	30	79.4	14	45	100
B	27	74.1	20	32	100

Table 6.10: Summary statistics of scores, split by exam version.

⊙ **Exercise 6.8** Construct a two-sided hypothesis test to evaluate whether the sample difference, $\bar{x}_A - \bar{x}_B = 5.3$, might be due to chance. [14]

⊙ **Exercise 6.9** To evaluate the hypotheses in Exercise 6.8 using the t distribution, we must first verify assumptions before moving forward with our methods. (a) Does it seem reasonable that the scores are independent? (b) What about the normality condition for each group? (c) Do you think each group would be independent? [15]

[13] This technique for degrees of freedom is conservative; it is more difficult to reject the null hypothesis using this *df* method.

[14] Because the professor did not expect one exam to be more difficult prior to examining the test results, she should use a two-sided hypothesis test. H_0: the exams are equally difficult, on average. $\mu_A - \mu_B = 0$. H_A: one exam was more difficult than the other, on average. $\mu_A - \mu_B \neq 0$.

[15] (a) It is probably reasonable to conclude the scores are independent. (b) The summary statistics suggest the data is roughly symmetric about the mean, and it doesn't seem unreasonable to suggest the data might be normal. (c) It seems reasonable to suppose that the samples are independent since the exams were handed out randomly.

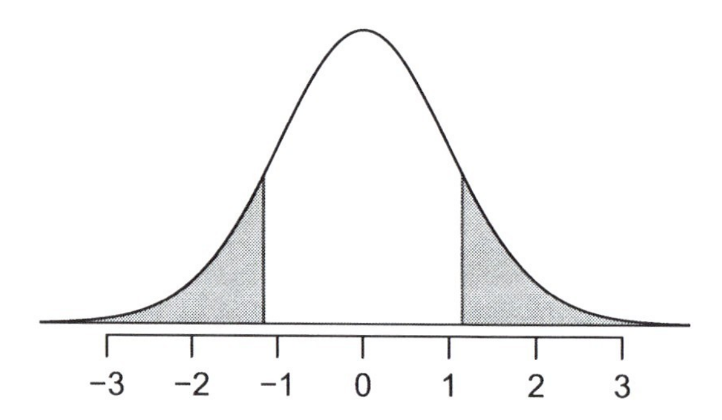

Figure 6.11: The t distribution with 26 degrees of freedom. The shaded right tail represents values with $T \geq 1.15$. Because it is a two-sided test, we also shade the corresponding lower tail.

With the conditions verified for each sample and the confirmation of independent samples, we move forward with the test and can safely utilize the t distribution. In this case, we are estimating the true difference in average test scores using the sample data, so the point estimate is $\bar{x}_A - \bar{x}_B = 5.3$. The standard error of the estimate can be identified using Equation (6.1):

$$SE = \sqrt{\frac{s_A^2}{n_A} + \frac{s_B^2}{n_B}} = \sqrt{\frac{14^2}{30} + \frac{20^2}{27}} = 4.62$$

Finally, we construct the test statistic:

$$T = \frac{\text{point estimate} - \text{null value}}{SE} = \frac{(79.4 - 74.1) - 0}{4.62} = 1.15$$

If we have a computer handy, we can identify the degrees of freedom as 45.97. Otherwise we use the smaller of $n_1 - 1$ and $n_2 - 1$: $df = 26$.

⊙ **Exercise 6.10** Identify the p-value, shown in Figure 6.11. Use $df = 26$. [16]

In Exercise 6.10, we could have used $df = 45.97$. However, this value is not listed in the table. In such cases, we use the next lower degrees of freedom (unless the computer also provides the p-value). For example, we could have used $df = 45$ but not $df = 46$.

⊙ **Exercise 6.11** Do embryonic stem cells (ESCs) help improve heart function following a heart attack? Table 6.13 contains summary statistics for an experiment

[16]We examine row $df = 26$ in the t table. Because this value is smaller than the value in the left column, the p-value is at least 0.200 (two tails!). Because the p-value is so large, we do not reject the null hypothesis. That is, the data does not convincingly show that one exam version is more difficult than the other, and the teacher is not convinced that she should add points to the Version B exam scores.

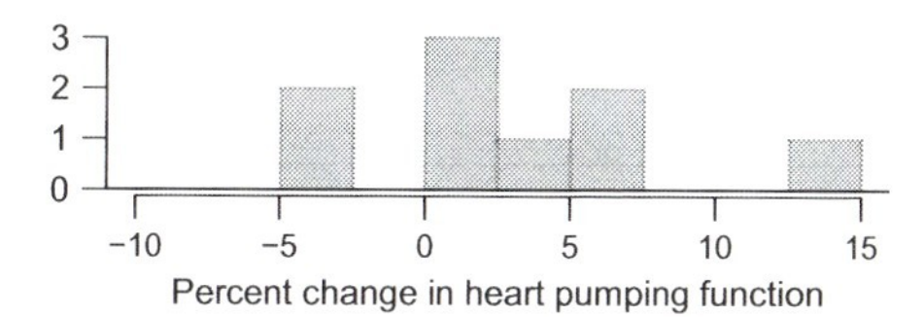

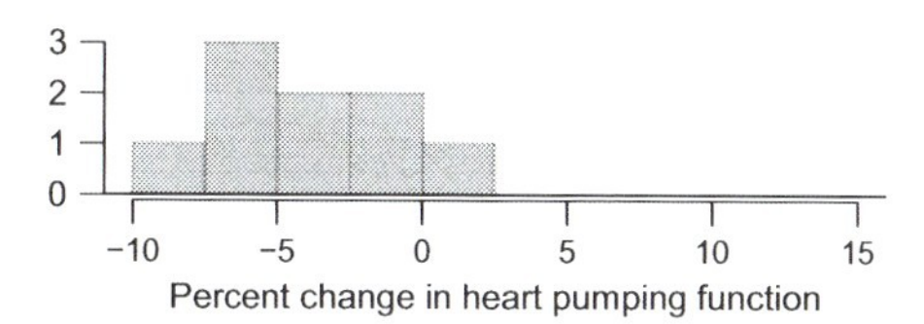

Figure 6.12: Histograms for both the embryonic stem cell group and the control group. Higher values are associated with greater improvement.

to test ESCs in sheep that had a heart attack. Each of these sheep was randomly assigned to the ESC or control group, and the change in their hearts' pumping capacity was measured. A positive value generally corresponds to increased pumping capacity, which suggests a stronger recovery. Setup hypotheses to test whether there is convincing evidence that ESCs actually increase the amount of blood the heart pumps. [17]

● **Example 6.5** The raw data from the ESC experiment described in Exercise 6.11 may be viewed in Figure 6.12. Using 8 degrees of freedom for the t distribution, evaluate the hypotheses.

We first compute the point estimate of the difference along with the standard error:

$$\bar{x}_{esc} - \bar{x}_{control} = 7.88$$

$$SE = \sqrt{\frac{5.17^2}{9} + \frac{2.76^2}{9}} = 1.95$$

The p-value is depicted as the shaded right tail in Figure 6.14. Computing the test

[17]We first setup the hypotheses:

H_0: The stem cells do not improve heart pumping function. $\mu_{esc} - \mu_{control} = 0$.

H_A: The stem cells do improve heart pumping function. $\mu_{esc} - \mu_{control} > 0$.

Before we move on, we must first verify that the t distribution method can be applied. Because the sheep were randomly assigned their treatment and, presumably, were kept separate from one another, the independence assumption is verified for each sample as well as for between samples. The data is very limited, so we can only check for obvious outliers in the raw data in Figure 6.12. Since the distributions are (very) roughly symmetric, we will assume the normality condition is acceptable. Because the conditions are satisfied, we can apply the t distribution.

	n	$\bar{x}$	s
ESCs	9	3.50	5.17
control	9	-4.33	2.76

Table 6.13: Summary statistics of scores, split by exam version.

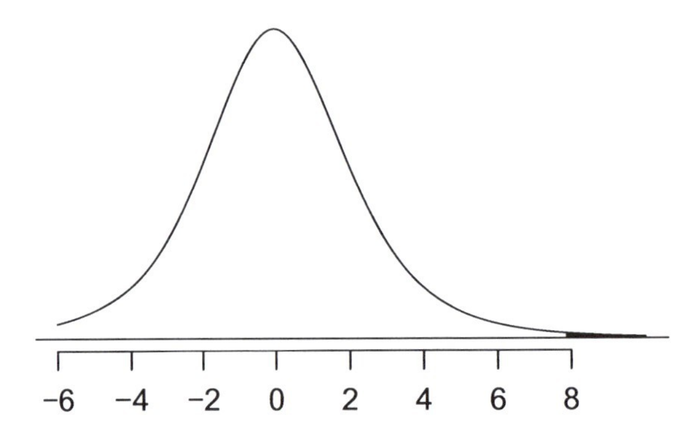

Figure 6.14: Distribution of the sample difference of the mean improvements if the null hypothesis was true. The shaded area represents the p-value.

statistic:

$$T = \frac{7.88 - 0}{1.95} = 4.03$$

We use the smaller of $n_1 - 1$ and $n_2 - 1$ (each are the same) for degrees of freedom: $df = 8$. Finally, we look for $T = 4.03$ in the t table; it falls to the right of the last column, so the p-value is smaller than 0.005 (one tail!). Because the p-value is less than 0.005 and therefore also smaller than 0.05, we reject the null hypothesis. The data provides convincing evidence that embryonic stem cells improve the heart's pumping function in sheep that have suffered a heart attack.

6.2.3 Two sample t confidence interval

Based on the results of Exercise 6.11, you found significant evidence that ESCs actually help improve the pumping function of the heart. But how large is this improvement? To answer this question, we can use a confidence interval.

⊙ **Exercise 6.12** In Exercise 6.11, you found that the point estimate, $\bar{x}_{esc} - \bar{x}_{control} = 7.88$, has a standard error of 1.95. Using $df = 8$, create a 99% confidence interval for the improvement due to ESCs. [18]

[18] We know the point estimate, 7.88, and the standard error, 1.95. We also verified the conditions for using the t distribution in Exercise 6.11. Thus, we only need identify t_8^* to create a 99% confidence interval: $t_8^* = 3.36$. Thus, the 99% confidence interval for the improvement from ESCs is given by

$$7.88 \pm 3.36 * 1.95 \quad \rightarrow \quad (1.33, 14.43)$$

That is, we are 99% confident that the true improvement in heart pumping function is somewhere between 1.33% and 14.43%.

6.3 Problem set

6.3.1 Small sample inference for the mean

6.1 An independent random sample is selected from normal population with unknown standard deviation. The sample is small ($n < 50$). Find the degrees of freedom and the critical t value (t*) for the given confidence level.

(a) $n = 42$, CL $= 90\%$

(b) $n = 21$, CL $= 98\%$

(c) $n = 29$, CL $= 95\%$

(d) $n = 12$, CL $= 99\%$

6.2 For a given confidence level, t^*_{df} is larger than z^*. Explain how t^*_{df} being slightly larger than z^* affects the width of the confidence interval.

6.3 An independent random sample is selected from normal population with unknown standard deviation. The sample is small ($n < 50$). Find the p-value for the given set of hypotheses and t values. Also determine if the null hypothesis would be rejected at $\alpha = 0.05$.

(a) $H_A : \mu > \mu_0$, $n = 11$, $T = 1.91$

(b) $H_A : \mu < \mu_0$, $n = 17$, $T = 3.45$

(c) $H_A : \mu \neq \mu_0$, $n = 38$, $T = 0.83$

(d) $H_A : \mu > \mu_0$, $n = 47$, $T = 2.89$

6.4 New York is known as "the city that never sleeps". A random sample of 25 New Yorkers were asked how much sleep they get per night. A summary of some sample statistics are shown below. Based on this sample, is there significant evidence to suggest that New Yorkers on average sleep less than 8 hrs a night?

n	$\bar{x}$	s	min	max
25	7.73	0.77	6.17	9.78

(a) Write the hypotheses in symbols and in words.

(b) Are the assumptions/conditions for inference satisfied?

(c) Calculate the test statistic, T.

(d) Find and interpret the p-value in context. Drawing a picture may be helpful.

(e) Based on the hypothesis test is there significant evidence to suggest that New Yorkers on average sleep less than 8 hrs a night?

(f) Would you expect a confidence interval for the population mean of at an equivalent confidence level as the hypothesis test to include 8? Explain.

6.5 Exercise 4 gives some summary statistics on the number of hours of sleep 25 randomly sampled New Yorkers get per night.

(a) Calculate a 90% confidence interval for the number of hours of New Yorkers sleep on average and interpret this interval in context.

(b) Does your confidence interval agree with the result of the hypothesis test from Exercise 4?

6.3.2 Using the t distribution for inference on the difference of two means

6.6 We are interested in comparing the average total personal income in Cleveland, OH and Sacramento, CA based on a random sample of individuals from the 2000 Census. Below are a histogram representing the distributions of total personal income of individuals living in Cleveland and Sacramento and some summary statistics on the two samples.

Is a t-test appropriate for testing whether or not there is a difference in the average incomes in these two metropolitan cities?

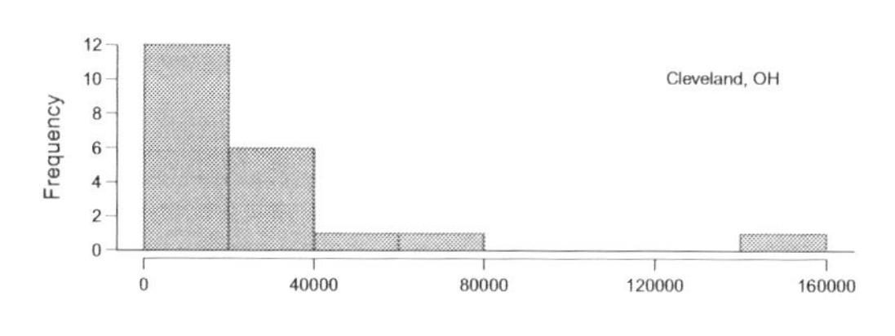

	Cleveland, OH
Mean	\$ 26,436
SD	\$ 33,239
n	21

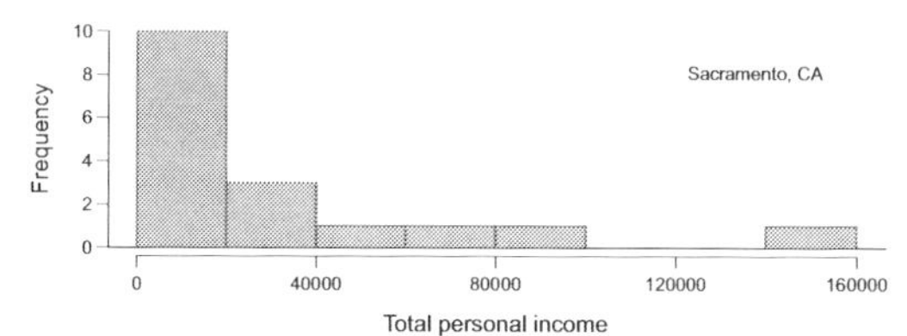

	Sacramento, CA
Mean	\$ 32,182
SD	\$ 40,480
n	17

6.7 Two independent random samples are selected from normal populations with unknown standard deviations. Both samples are small ($n < 50$). Find the p-value for the given set of hypotheses and t values. Also determine if the null hypothesis would be rejected at $\alpha = 0.05$.

(a) $H_A : \mu_1 > \mu_2$, $n_1 = 23$, $n_2 = 25$, $T = 3.16$

(b) $H_A : \mu_1 \neq \mu_2$, $n_1 = 38$, $n_2 = 37$, $T = 2.72$

(c) $H_A : \mu_1 < \mu_2$, $n_1 = 45$, $n_2 = 41$, $T = 1.83$

(d) $H_A : \mu_1 \neq \mu_2$, $n_1 = 11$, $n_2 = 15$, $T = 0.28$

6.8 Two independent random samples are selected from normal populations with unknown standard deviations. Both samples are small ($n < 50$). Find the degrees of freedom and the critical t value (t^*) for the given confidence level. Remember that a reasonable choice of degrees of freedom for the two-sample case is the minimum of $n_1 - 1$ and $n_2 - 1$. The "exact" df is something we cannot compute from the given info.

(a) $n_1 = 16$, $n_2 = 16$, CL = 90%

(b) $n_1 = 36$, $n_2 = 41$, CL = 95%

(c) $n_1 = 8$, $n_2 = 10$, CL = 99%

(d) $n_1 = 23$, $n_2 = 27$, CL = 98%

6.9 A weight loss pill claims to accelerate weight loss when accompanied with exercise and diet. Diet researchers from a consumer advocacy group decided to test this claim using an experiment. 42 subjects were randomly assigned to two groups: 21 took the pill and 21 only received a placebo. Both groups underwent the same diet and exercise regiment. In the group that got the pill the average weight loss was $20lbs$ with a standard deviation of $4lbs$. In the placebo group the average weight loss was $18lbs$ with a standard deviation of $5lbs$.

(a) Calculate a 95% confidence interval for the difference between the two means.

(b) Interpret the confidence interval in context.

(c) Based on your confidence interval, is there significant evidence to suggest that the weight loss pill is effective?

(d) Does this prove that the weight loss pill is/is not effective?

(a) Before we can calculate the confidence interval, we need to find the degrees of freedom and t^*_{df}.

$$df = min(n_1 - 1, n_2 - 1) = min(20, 20) = 20$$
$$t^*_{20} = 2.086$$

$$\begin{aligned}(\bar{x}_1 - \bar{x}_2) \pm t^*_{df}\sqrt{\frac{s_1^2}{n_1} + \frac{s_2^2}{n_2}} &= (20 - 18) \pm 2.086 * \sqrt{\frac{4^2}{21} + \frac{5^2}{21}} \\ &= 2 \pm 2.91 \\ &= (-0.91, 4.91)\end{aligned}$$

(b) We are 95% confident that those in the group that got the weight loss pill lost $0.91lbs$ less to $4.91lbs$ more than those in the placebo group.

(c) Since the confidence interval includes 0 there is no significant evidence to suggest that the weight loss pill is effective.

(d) No, we can only say that the data *suggest* that the pill is not effective, we did not prove that it is not. There may be other contributing factors.

6.10 A company has two factories in which they manufacture engines. Once a month they randomly select 10 engines from each factory to test if there is a difference in performance in engines made in the two factories. The average output of the motors from Factory 1 is 120 horsepower with a standard deviation of 5 horsepower. The average output of the motors from Factory 2 is 132 horsepower with a standard deviation of 4 horsepower.

(a) Calculate a 95% confidence interval for the difference in the average horsepower for engines coming from the two factories.

(b) Interpret the confidence interval in context.

(c) Based on your confidence interval, is there significant evidence to suggest that there is a difference in performance in engines made in the two factories? If so, can you tell which factory produces motors with lower performance? Explain.

(d) Recently upgrades were made in Factory 2. Does this prove that these upgrades enhanced the performance in engines made in the this factory? Explain.

6.11 An experiment was conducted to measure and compare the effectiveness of various feed supplements on the growth rate of chickens. Newly hatched chicks were randomly allocated into six groups, and each group was given a different feed supplement. Their weights in grams after six weeks are given along with feed types in the data set called `chickwts` [13]. Below are some summary statistics from this data set along with box plots showing the distribution of weights by feed type.

	Mean	SD	n
casein	323.58	64.43	12
horsebean	160.20	38.63	10
linseed	218.75	52.24	12
meatmeal	276.91	64.90	11
soybean	246.43	54.13	14
sunflower	328.92	48.84	12

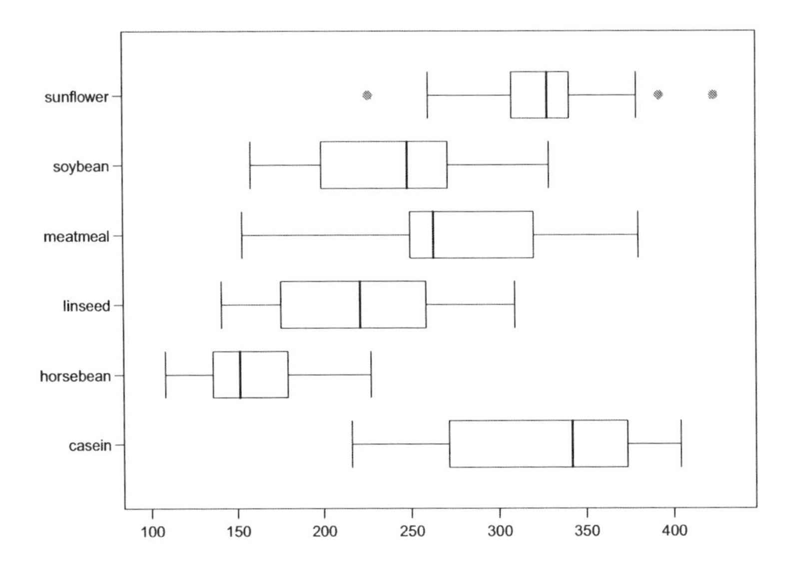

(a) Describe the distributions of weights of chicken that were fed linseed and horsebean.

(b) Write the hypotheses for testing for a significant difference between the average weights of chicken that were fed linseed and horsebean.

(c) Are the assumptions and conditions for inference satisfied?

(d) Calculate the test statistic and find the p-value.

(e) Using $\alpha = 0.05$ what do you conclude? Interpret your conclusion in context.

(f) What type of error might we have committed? Explain.

(g) Would your conclusion change if we used $\alpha = 0.01$?

6.12 Casein is a common weight gain supplement for humans. Does it have the same effect on chicken? Using data provided in Exercise 11 test the hypothesis that the average weight of chickens that were fed casein is higher than the average weight of chickens that were fed soybean. Assume that conditions for inference are satisfied and the data are symmetricly distributed.

6.13 Each year the US Environmental Protection Agency (EPA) releases fuel economy data on cars manufactured in that year [14]. Below are some summary statistics on fuel efficiency (in miles/gallon) random samples of cars with manual and automatic transmission cars manufactured in 2010. Based on this information, is there a significant difference between the average fuel efficiency of cars with manual and automatic transmission in terms of their city mileage? Assume that conditions for inference are satisfied and the data are symmetricly distributed.

	Mean City MPG	SD City MPG	Mean Hwy MPG	SD Hwy MPG	n
Manual	21.08	4.29	29.31	4.63	26
Automatic	15.62	2.76	21.38	3.73	26

6.14 Exercise 13 provides data on fuel efficiency of cars manufactured in 2010. Based on this information, calculate a 95% confidence interval for the difference between average highway mileage of manual and automatic cars and interpret this interval in context.

Chapter 7

Introduction to linear regression

Linear regression is a very powerful statistical technique. Many people have some familiarity with regression just from reading the news, where graphs with straight lines are overlaid on scatterplots. Linear models can be used for prediction or to evaluate whether there is a linear relationship between two numerical variables.

Figure 7.1 shows two variables that can be modeled perfectly with a straight line. The equation for the line is

$$y = -5 + 1.35x$$

Imagine what a perfect linear relationship would mean: you would know the exact value of y, just by knowing the value of x. This is unrealistic in almost any natural process. Consider height and weight of school children for example. Their height, x, gives you some information about their weight, y, but there is still a lot of variability, even for children of the same height.

We often write a linear regression line as

$$y = b_0 + b_1 x$$

where b_0 and b_1 represent two parameters that we wish to identify. Usually x represents an explanatory or *predictor* variable and y represents a response. We use the variable x to predict a response y.

Examples of several scatterplots are shown in Figure 7.2. While none reflect a perfect linear relationship, it will be useful to fit approximate linear relationships to each. This line represents a model relating x to y. The first plot shows a relatively strong downward linear trend. The second shows an upward trend that, while evident, is not as strong as the first. The last plot shows a very weak downward trend in the data, so slight we can hardly notice it. In Section 7.4 (this section is not included in this textbook version), we develop a way to quantify how well those "best lines" fit the data.

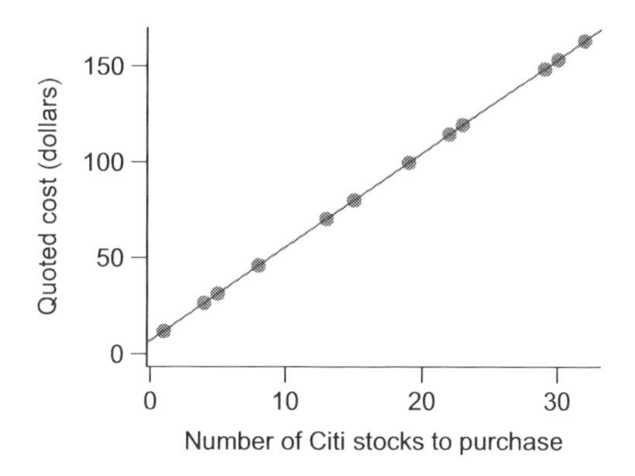

Figure 7.1: Twelve requests were put into a trading company to buy Citi stock (ticker C), and the cost quotes were reported (April 22nd, 2010). Because the cost is computed using a linear formula, the linear fit is perfect.

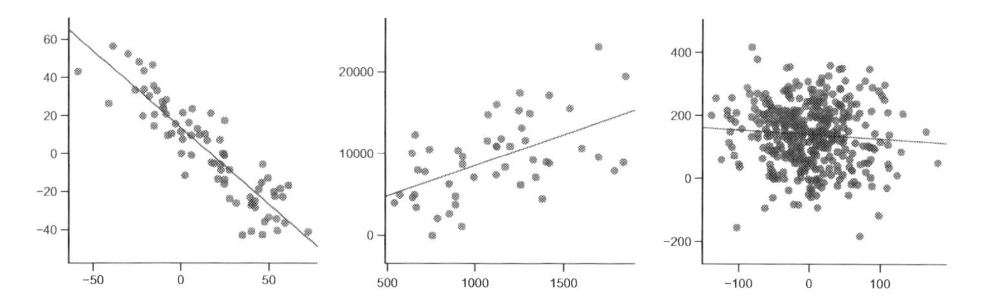

Figure 7.2: Three data sets where a linear model may be useful but is not perfect.

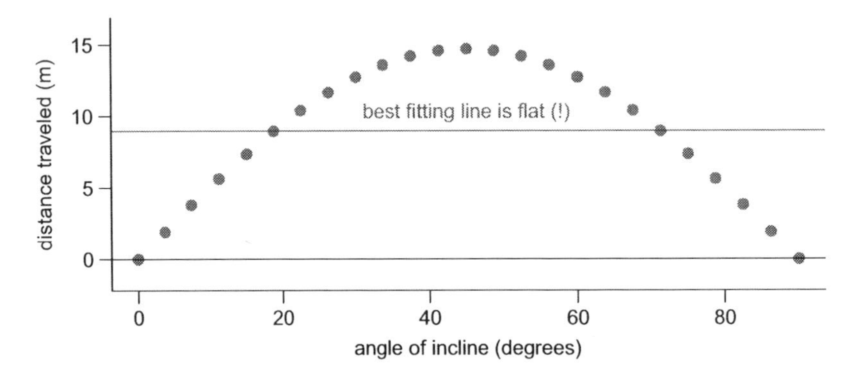

Figure 7.3: A linear model is not useful in this non-linear case. (This data is from an introductory physics experiment.)

We will soon find that there are cases where fitting a straight line to the data, even if there is a clear relationship between the variables, is not helpful. One such case is shown in Figure 7.3 where there is a very strong relationship between the variables even if the trend is not linear. We tackle such non-linear trends in Chapter 8.

7.1 Line fitting, residuals, and correlation

It is helpful to think deeply about the line fitting process. In this section, we examine helpful criteria for identifying a linear model and introduce a new statistic, *correlation*.

7.1.1 Beginning with straight lines

Scatterplots were introduced in Chapter 1 as a graphical technique to present two numerical variables simultaneously. Such plots permit the relationship between the variables to be examined with ease. Figure 7.4 shows a scatterplot for the `headL` and `totalL` variables from the `possum` data set. Each point represents a single possum from the data.

The `headL` and `totalL` variables are associated. Possums with an above average total length also tend to have above average head lengths. While the relationship is not perfectly linear, it could be helpful to partially explain the connection between these variables with a straight line.

Straight lines should only be used when the data appear to have a linear relationship, such as the case shown in the left panel of Figure 7.5. The right panel of Figure 7.5 shows a case where a curved band would be more useful in capturing a different set of data. We will discuss how to handle non-linear relationships in Chapter 8.

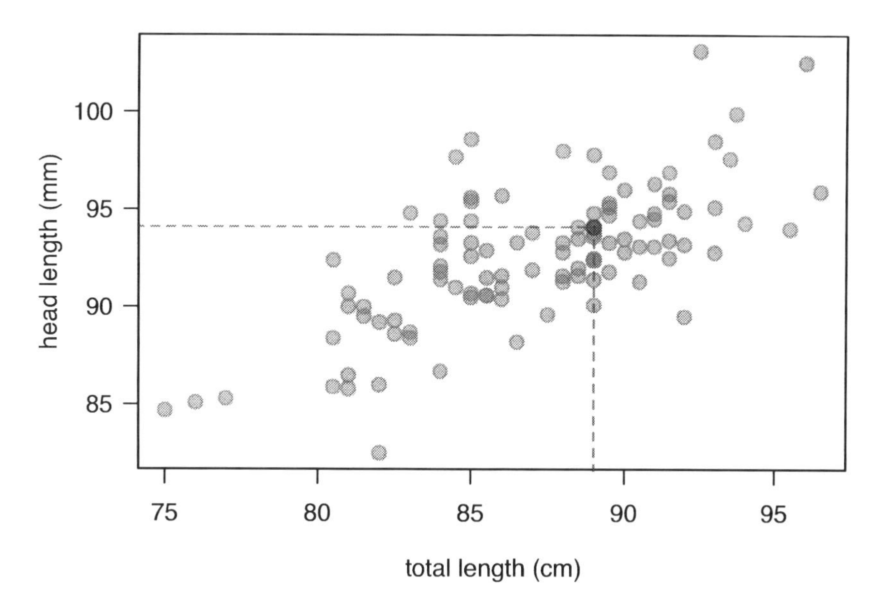

Figure 7.4: A scatterplot showing `headL` against `totalL`. The first possum with a head length of 94.1mm and a length of 89cm is highlighted.

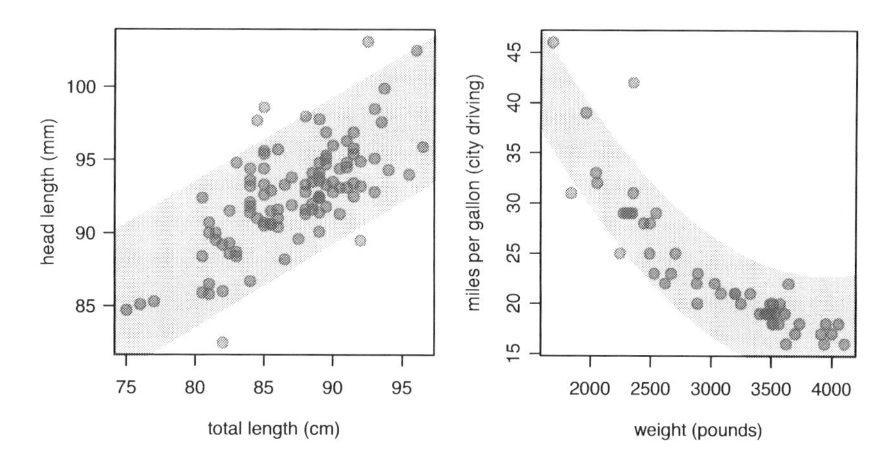

Figure 7.5: Most observations on the left can be captured in a straight band. On the right, we have the `weight` and `mpgCity` variables represented, and a curved band does a better job of capturing these cases than a straight band.

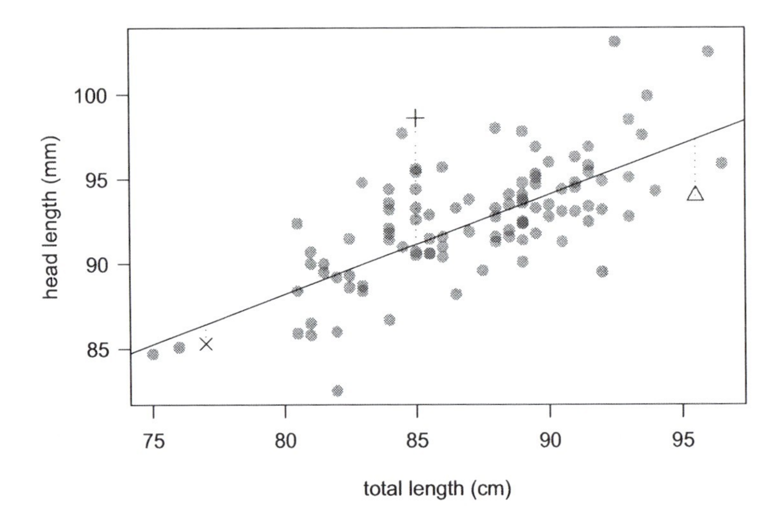

Figure 7.6: A reasonable linear model was fit to represent the relationship between `headL` and `totalL`.

> **Caution: Watch out for curved trends**
> We only consider models based on straight lines in this chapter. If data shows a non-linear trend, like that in the right panel of Figure 7.5, more advanced techniques should be used.

7.1.2 Fitting a line by eye

We want to describe the relationship between the `headL` and `totalL` variables using a line. In this example, we will use the total length as the predictor variable, x, to predict a possum's head length, y.

We could fit the linear relationship by eye, as in Figure 7.6. The equation for this line is

$$\widehat{y} = 41 + 0.59 * x \tag{7.1}$$

We can use this model to discuss properties of possums. For instance, our model predicts a possum with a total length of 80 cm will have a head length of

$$\widehat{y} = 41 + 0.59 * 80 = 88.2mm$$

A "hat" on y is used to signify that this is an estimate. Linear models predict the average value of y for a particular value of x based on the model. For example, the model predicts that possums with a total length of 80 cm will have an average head length of 88.2 mm. Without further information about an 80 cm possum, this prediction for head length that uses the average is a reasonable estimate.

7.1.3 Residuals

Residuals can be thought of as the leftovers from the model fit:

$$\text{Result} = \text{Fit} + \text{Residual}$$

Each observation will have a residual. If an observation is above the regression line, then its residual, the vertical distance from the observation to the line, is positive. Observations below the line have negative residuals. One goal in picking the right linear model is for these residuals to be as small as possible.

Three observations are noted specially in Figure 7.6. The "×" has a small, negative residual of about -1, the observation marked by "+" has a large residual of about +7, and the observation marked by "△" has a moderate residual of about -4. The size of residuals is usually discussed in terms of absolute value. For example, the residual for △ is larger than that of × because $|-4|$ is larger than $|-1|$.

Residual: difference of observed and expected
The *residual* of an observation (x_i, y_i) is the difference of the observed response (y_i) and the response we would predict based on the model fit ($\hat{y}_i$):

$$r_i = y_i - \hat{y}_i$$

We typically identify $\hat{y}_i$ by plugging x_i into the model. The residual of the ith observation is denoted by r_i.

● **Example 7.1** The linear fit shown in Figure 7.6 is given in Equation (7.1). Based on this line, formally compute the residual of the observation (77.0, 85.3). This observation is denoted by × on the plot. Check it against the earlier visual estimate, -1.

We first compute the predicted value based on the model:

$$\hat{y}_\times = 41 + 0.59 * x_\times = 41 + 0.59 * 77.0 = 86.4$$

Next we compute the difference of the actual head length and the predicted head length:

$$r_\times = y_\times - \hat{y}_\times = 85.3 - 86.43 = -0.93$$

This is very close to the visual estimate of -1.

⊙ **Exercise 7.1** If a model underestimates an observation, will the residual be positive or negative? What about if it overestimates the observation? [1]

[1]If a model underestimates an observation, then the model estimate is below the actual. The residual – the actual minus the model estimate – must then be positive. The opposite is true when the model overestimates the observation: the residual is negative.

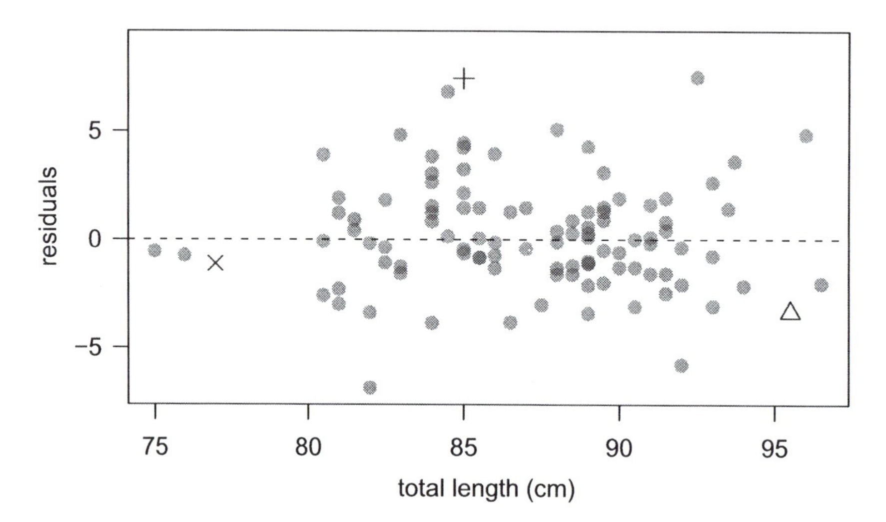

Figure 7.7: Residual plots for the two models in Figure 7.6.

⊙ **Exercise 7.2** Compute the residuals for the observations (85.0, 98.6) (+ in the figure) and (95.5, 94.0) (△) using the linear model given in Equation (7.1). Answer for + is in the footnote[2].

Residuals are helpful in evaluating how well a linear model fits a data set. We often plot them in a **residual plot** such as the one shown in Figure 7.7 for the line in Figure 7.6. The residuals are plotted at their original horizontal locations but with the vertical coordinate as the residual. For instance, the point $(85.0, 98.6)_+$ had a residual of 7.45, so in the residual plot it is placed at (85.0, 7.45). Creating a residual plot is sort of like tipping the scatterplot over so the regression line is horizontal.

● **Example 7.2** One purpose of residual plots is to identify characteristics or patterns still apparent in data after fitting a model. Figure 7.8 shows three scatterplots with linear models in the first row and residual plots in the second row. Can you identify any patterns remaining in the residuals?

In the first data set (first column), the residuals show no obvious patterns. The residuals appear to be scattered randomly about 0, represented by the dashed line.

The second data set shows a pattern in the residuals. There is some curvature in the scatterplot, which is more obvious in the residual plot. We should not use a straight

[2]First compute the predicted value based on the model:

$$\hat{y}_+ = 41 + 0.59 * x_+ = 41 + 0.59 * 85.0 = 91.15$$

Then the residual is given by

$$r_+ = y_+ - \hat{y}_+ = 98.6 - 91.15 = 7.45$$

This was close to the earlier estimate of 7.

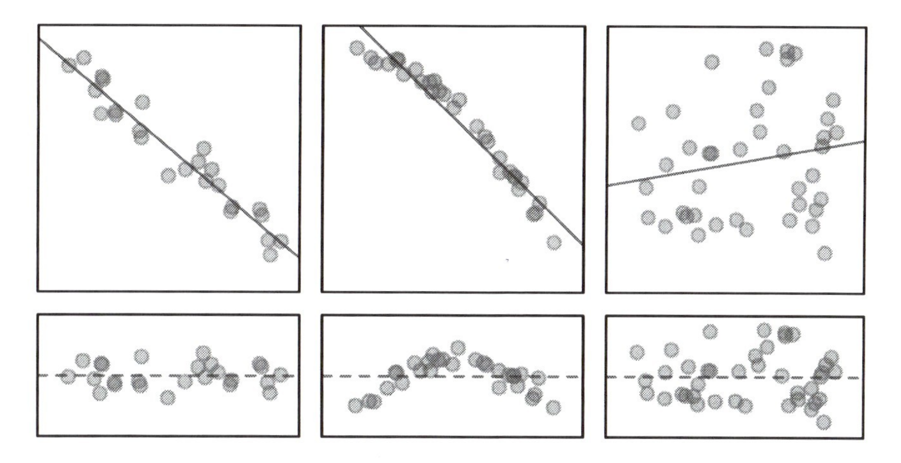

Figure 7.8: Sample data with their best fitting lines (top row) and their corresponding residual plots (bottom row).

line to model this data but should use a more advanced technique from Chapter 8.

The last plot shows very little upwards trend, and the residuals also show no obvious patterns. It is reasonable to try to fit this linear model to the data, however, it is unclear whether this model will be significant. That is, could we have seen this slight trend just due to chance? We will tackle this sort of question in Section 7.4 (this section is not included in this textbook version).

7.1.4 Describing linear relationships with correlation

Correlation is useful in describing how linear the relationship is between two variables.

> **Correlation: strength of a linear relationship**
> The **correlation** describes the strength of the linear relationship between two variables and takes values between -1 and 1. We denote the correlation by R.

We compute the correlation using a formula, just as we did with the sample mean and standard deviation. However, this formula is rather complex[3], so we generally perform the calculations on a computer or calculator. Figure 7.9 shows eight plots and their corresponding correlations. Only when the relationship is perfectly linear is the correlation either -1 or 1. If the relationship is strong and

[3]Formally, we can compute the correlation for observations (x_1, y_1), (x_2, y_2), ..., (x_n, y_n) using the formula

$$R = \frac{1}{s_x s_y} \sum_{i=1}^{n} (x_i - \bar{x})(y_i - \bar{y})$$

where $\bar{x}$, $\bar{y}$, s_x, and s_y are the sample means and standard deviations for each variable.

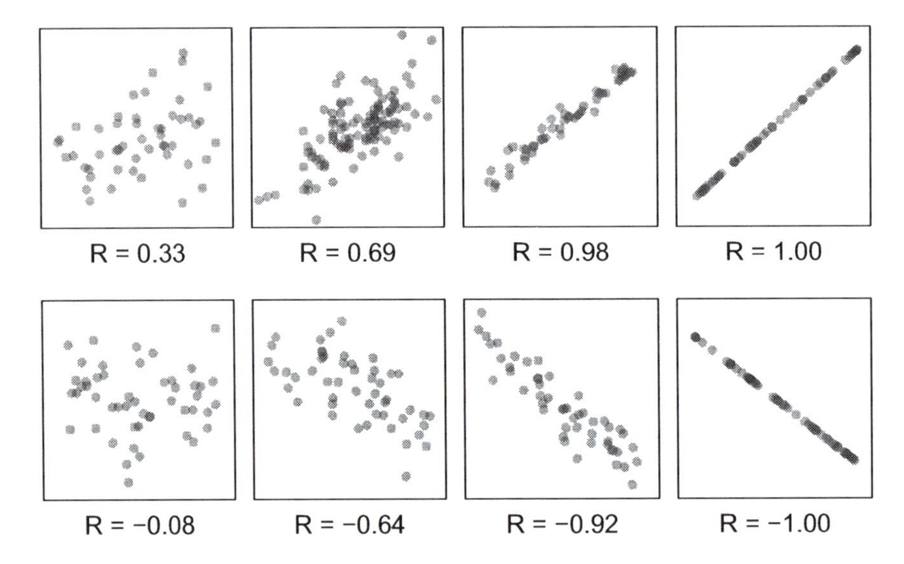

Figure 7.9: Sample scatterplots and their correlations. The first row shows variables with a positive relationship, represented by the trend up and to the right. The second row shows variables with a negative trend, where a large value in one variable is associated with a low value in the other.

positive, the correlation will be near +1. If it is strong and negative, it will be near -1. If there is no apparent linear relationship between the variables, then the correlation will be near zero.

The correlation is intended to quantify linear trends. Non-linear trends, even when strong, sometimes produce correlations that do not reflect the strength of the relationship; see three such examples in Figure 7.10.

⊙ **Exercise 7.3** While there is not a linear model that fits each of the curves in Figure 7.10, try fitting a non-linear curve to each figure. Once you create a curve for each, describe what is important in your fit. Why did you make the lines the way

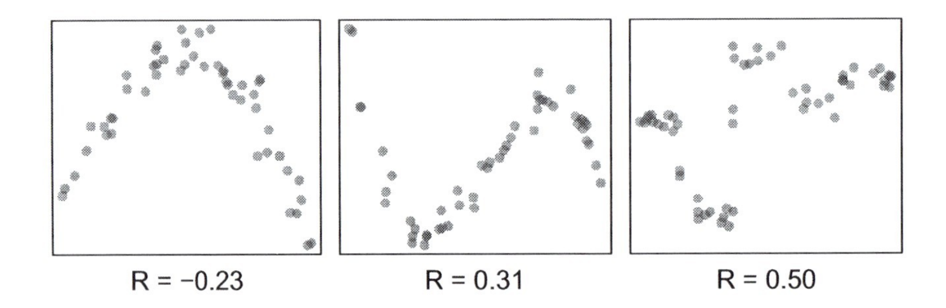

Figure 7.10: Sample scatterplots and their correlations. In each case, there is a strong relationship between the variables. However, the correlation is not very strong.

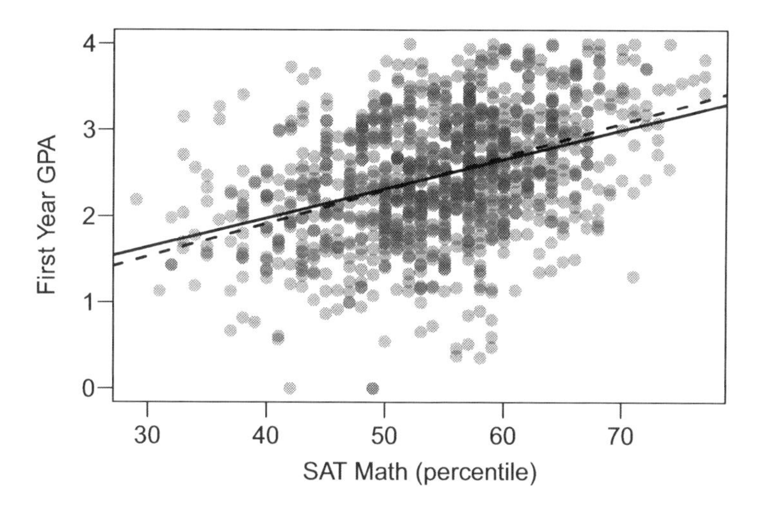

Figure 7.11: SAT math (percentile) and first year GPA scores. Two lines are fit to the data, the solid line being the *least squares line*.

you did? [4]

7.2 Fitting a line by least squares regression

Fitting linear models by eye is (rightfully) open to criticism since it is based on an individual preference. In this section, we define a criteria to help us describe what is a good fitting line, which leads to a commonly used model fitting technique called *least squares regression*.

This section will use SAT math scores and first year GPA for a random sample of students at a college[5]. A scatterplot of the data is shown in Figure 7.11 along with two linear fits. The lines follow a positive trend in the data; students who scored higher math SAT scores also tended to have higher GPAs after their first year in school.

⊙ **Exercise 7.4** Is the correlation positive or negative? [6]

7.2.1 An objective measure for finding the best line

We begin by thinking about what we mean by "best". Mathematically, we want a line that has small residuals. Perhaps our criteria could minimize the sum of the

[4]Possible explanation: The line should be close to most points and reflect trends in the data.

[5]This data was collected by Educational Testing Service from an unnamed college.

[6]Positive: because larger SAT scores are associated with higher GPAs, the correlation will be positive. Using a computer (or by doing a lot of hand computations), the correlation can be computed: 0.387.

residual magnitudes:

$$|r_1| + |r_2| + \cdots + |r_n| \tag{7.2}$$

We could program a computer to find a line that minimizes this criteria (the sum). This does result in a pretty good fit, which is shown as the dashed line in Figure 7.11. However, a more common practice is to choose the line that minimizes the sum of the squared residuals:

$$r_1^2 + r_2^2 + \dots r_n^2 \tag{7.3}$$

The line that minimizes this **least squares criteria** is represented as the solid line in Figure 7.11. This is commonly called the **least squares line**.

Three possible reasons to choose Criteria (7.3) over Criteria (7.2) are the following:

1. It is the most commonly used method.
2. Computing the line based on Criteria (7.3) is much easier by hand and in most statistical softwares.
3. In many applications, a residual twice as large as another is more than twice as bad. For example, being off by 4 is usually more than twice as bad as being off by 2. Squaring the residuals accounts for this discrepancy.

The first two reasons are largely for tradition and convenience, and the last reason explains why Criteria (7.3) is typically most helpful[7].

7.2.2 Conditions for the least squares line

When fitting a least squares line, we generally require

- **Linearity.** The data should show a linear trend. If there is a non-linear trend (e.g. left panel of Figure 7.12), an advanced method from Chapter 8 should be applied.
- **Nearly normal residuals.** Generally the residuals must be nearly normal. When this condition is found to be unreasonable, it is usually because of outliers or concerns about influential points, which we will discuss in greater depth in Section 7.3. An example of non-normal residuals is shown in the center panel of Figure 7.12.
- **Constant variability.** The variability of points around the least squares line remains roughly constant. An example of non-constant variability is shown in the right panel of Figure 7.12.

[7]There are applications where Criteria (7.2) may be more useful, and there are plenty of other criteria we might consider. However, this course only applies the least squares criteria for regression.

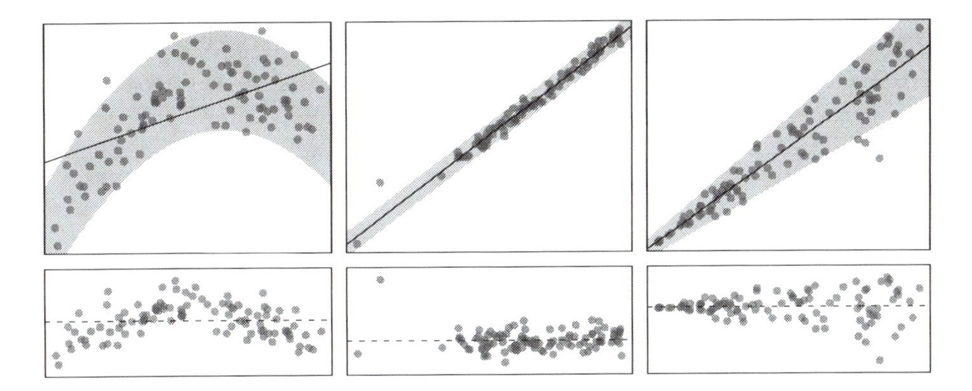

Figure 7.12: Three examples showing when not to use simple least squares regression. In the left panel, a straight line does not fit the data. In the second panel, there are outliers; two points on the left are relatively distant from the rest of the data and one of these points is very far away from the line. In the third panel, the variability of the data around the line gets larger with larger values of x.

⊙ **Exercise 7.5** Do you have any concerns about applying least squares to the `satMath` and `GPA` data in Figure 7.11 on page 231?[8]

7.2.3 Finding the least squares line

For the `satMath` and `GPA` data, we could write the equation as

$$\widehat{GPA} = b_0 + b_1 * satMath \tag{7.4}$$

Here the equation is set up to predict `GPA` based on a student's `satMath` score, which would be useful to a college admissions office. These two values, b_0 and b_1, are the *parameters* of the regression line.

Just like in Chapters 4-6, the parameters are estimated using observed data. In practice, this estimation is done using a computer in the same way that other estimates, like a sample mean, can be estimated using a computer or calculator. However, we can also find the parameter estimates by applying two properties of the least squares line[9]:

- If $\bar{x}$ is the mean of the horizontal variable (from the data) and $\bar{y}$ is the mean of the vertical variable, then the point $(\bar{x}, \bar{y})$ is on the least squares line.
- The slope of the least squares line is estimated by

$$\hat{b}_1 = \frac{s_y}{s_x} R \tag{7.5}$$

[8]The trend appears to be linear, the data falls around the line with no obvious outliers, and the variance is roughly constant. Least squares regression can be applied to this data.

[9]These properties follow from the mathematics that lie behind least squares regression. Deriving these properties is beyond the scope of this course.

where R is the correlation between the two variables, and s_x and s_y are the sample standard deviations of the explanatory variable (variable on the horizontal axis) and response (variable on the vertical axis), respectively.

⊙ **Exercise 7.6** Table 7.13 shows the sample means for the `satMath` and `GPA` variables: 54.395 and 2.468. Plot the point $(54.395, 2.468)$ on Figure 7.11 on page 231 to verify it falls on the least squares line (the solid line).

	`satMath` ("x")	`GPA` ("y")
mean	$\bar{x} = 54.395$	$\bar{y} = 2.468$
sd	$s_x = 8.450$	$s_y = 0.741$
	correlation: $R = 0.387$	

Table 7.13: Summary statistics for the `satMath` and `GPA` data.

⊙ **Exercise 7.7** Using the summary statistics in Table 7.13, compute the slope for Equation (7.4). [10]

You might recall from math class the **point-slope** form of a line (another common form is *slope-intercept*). Given the slope of a line and a point on the line, (x_0, y_0), the equation for the line can be written as

$$y - y_0 = slope * (x - x_0) \tag{7.6}$$

A common exercise to get familiar with foundations of least squares regression is to use basic summary statistics and point-slope form to produce the least squares line.

TIP: Identifying the least squares line from summary statistics
To identify the least squares line from summary statistics,

- estimate the slope parameter, $\hat{b}_1$, using Equation (7.5),
- note that the point $(\bar{x}, \bar{y})$ is on the least squares line, use $x_0 = \bar{x}$ and $y_0 = \bar{y}$ along with the slope $\hat{b}_1$ in the point-slope equation:

$$y - \bar{y} = \hat{b}_1(x - \bar{x})$$

- simplify the equation.

[10] Apply Equation (7.5) with the summary statistics from Table 7.13 to compute the slope:

$$\hat{b}_1 = \frac{s_y}{s_x}R = \frac{0.741}{8.450}0.387 = 0.03394$$

Example 7.3 Using the point (54.395, 2.468) from the sample means and the slope estimate $\hat{b}_1 = 0.034$ from Exercise 7.7, find the least-squares line for predicting GPA based on satMath.

Apply the point-slope equation using (54.395, 2.468) and the slope, $\hat{b}_1 = 0.03394$:

$$y - y_0 = \hat{b}_1(x - x_0)$$
$$y - 2.468 = 0.03394(x - 54.395)$$

Multiplying out the right side and then adding 2.468 to each side, we find the least squares line:

$$\widehat{GPA} = 0.622 + 0.03394 * satMath$$

Here we have replaced y with $\widehat{GPA}$ and x with $satMath$ to put the equation in context. This form matches the form of Equation (7.4).

We mentioned earlier that a computer is usually used to compute the least squares line. A summary table based on some computer output is shown in Table 7.14 for the satMath and GPA data. The first column of numbers provide estimates for b_0 and b_1, respectively. Compare these to the result from Example 7.3.

	Estimate	Std. Error	t value	Pr(>|t|)
(Intercept)	0.6219	0.1408	4.42	0.0000
satMath	0.0339	0.0026	13.26	0.0000

Table 7.14: Summary of least squares fit for the SAT/GPA data. Compare the parameter estimates in the first column to the results of Example 7.3.

Exercise 7.8 Examine the second, third, and fourth columns in Table 7.14. Can you guess what they represent? [11]

7.2.4 Extrapolation is treacherous

When those blizzards hit the East Coast this winter, it proved to my satisfaction that global warming was a fraud. That snow was freezing cold. But in an alarming trend, temperatures this spring have risen. Consider this: On February 6th it was 10 degrees. Today it hit almost 80. At this rate, by August it will be 220 degrees. So clearly folks the climate debate rages on.

Stephen Colbert
April 6th, 2010 [12]

[11] These columns help us determine whether the estimates are significantly different from zero. The second column lists the standard errors of the estimates, the third column are t test statistics, and the fourth column lists p-values (2-sided test). We will describe the interpretation of these columns in greater detail in Section 7.4 (this section is not included in this textbook version).

[12] http://www.colbertnation.com/the-colbert-report-videos/269929/

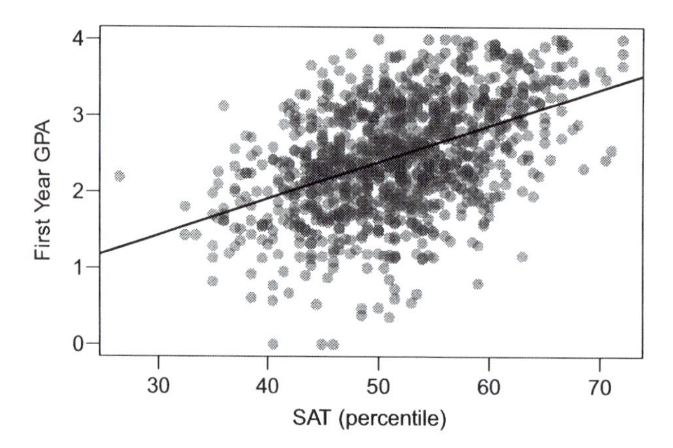

Figure 7.15: GPA scores and overall SAT percentile. The least squares regression line is shown.

Linear models are used to approximate the relationship between two variables, and these models have real limitations. Linear regression is simply a modeling framework. The truth is almost always much more complex than our simple line. For example, we do not know how the data outside of our limited window will behave.

SAT math scores were used to predict freshman GPA in Section 7.2.3. We could also use the overall percentile as a predictor in place of just the math score:

$$\widehat{GPA} = 0.0019 + 0.0477 * satTotal$$

These data are shown in Figure 7.15. The linear model in this case was built for observations between the 26th and the 72nd percentiles. The data meets all conditions necessary for the least squares regression line, so could the model safely be applied to students at the 90th percentile?

⊙ **Exercise 7.9** Use the model $\widehat{GPA} = 0.0019 + 0.0477 * satTotal$ to estimate the GPA of a student who is at the 90th percentile. [13]

Applying a model estimate to values outside of the realm of the original data is called **extrapolation**. Generally, a linear model is only an approximation of the real relationship between two variables. If we extrapolate, we are making an unreliable bet that the approximate linear relationship will be valid in places where it has not been evaluated.

[13]Predicting GPA for a person with an SAT percentile of 90:

$$0.0019 + 0.0477 * satTotal = 0.0019 + 0.0477 * 90 = 4.29$$

The model predicts a GPA score of 4.29. GPAs only go up to 4.0 (!).

7.2.5 Using R^2 to describe the strength of a fit

We evaluated the strength of the linear relationship between two variables earlier using correlation, R. However, it is more common to explain the strength of a linear fit using R^2, called **R-squared**. If we are given a linear model, we would like to describe how closely the data clusters around the linear fit.

The R^2 of a linear model describes the amount of variation in the response that is explained by the least squares line. For example, consider the SAT-GPA data, shown in Figure 7.15. The variance of the response variable, `GPA`, is $s^2_{GPA} = 0.549$. However, if we apply our least squares line, then this model reduces our uncertainty in predicting GPA using a student's SAT score. The variability in the residuals describes how much variation remains after using the model: $s^2_{RES} = 0.433$. In short, there was a reduction of

$$\frac{s^2_{GPA} - s^2_{RES}}{s^2_{GPA}} = \frac{0.549 - 0.433}{0.549} = \frac{0.116}{0.549} = 0.21$$

or about 21% in the data's variation by using information about the SAT scores via a linear model. This corresponds exactly to the R-squared value:

$$R = 0.46 \qquad\qquad R^2 = 0.21$$

⊙ **Exercise 7.10** If a linear model has a very strong negative relationship with a correlation of -0.97, how much of the variation in the response is explained by the explanatory variable? [14]

7.3 Types of outliers in linear regression

In this section, we identify (loose) criteria for which outliers are important and influential.

Outliers in regression are observations that fall far from the "cloud" of points. These points are especially important because they can have a strong influence on the least squares line.

⊙ **Exercise 7.11** There are six plots shown in Figure 7.16 along with the least squares line. For each plot, identify any obvious outliers and note how you think they influence the least squares line. Recall that an outlier is any point that doesn't appear to belong with the vast majority of the other points. [15]

[14] About $R^2 = (-0.97)^2 = 0.94$ or 94% of the variation is explained by the linear model.

[15] Across the top, then across the bottom: (1) There is one outlier far from the other points, though it only appears to slightly influence the line. (2) One outlier on the right, though it is quite close to the least squares line, which suggests it wasn't very influential. (3) One point is far away from the cloud, and this outlier appears to pull the least squares line up on the right; examine how the line around the primary cloud doesn't appear to fit very well. (4) There is a primary cloud and

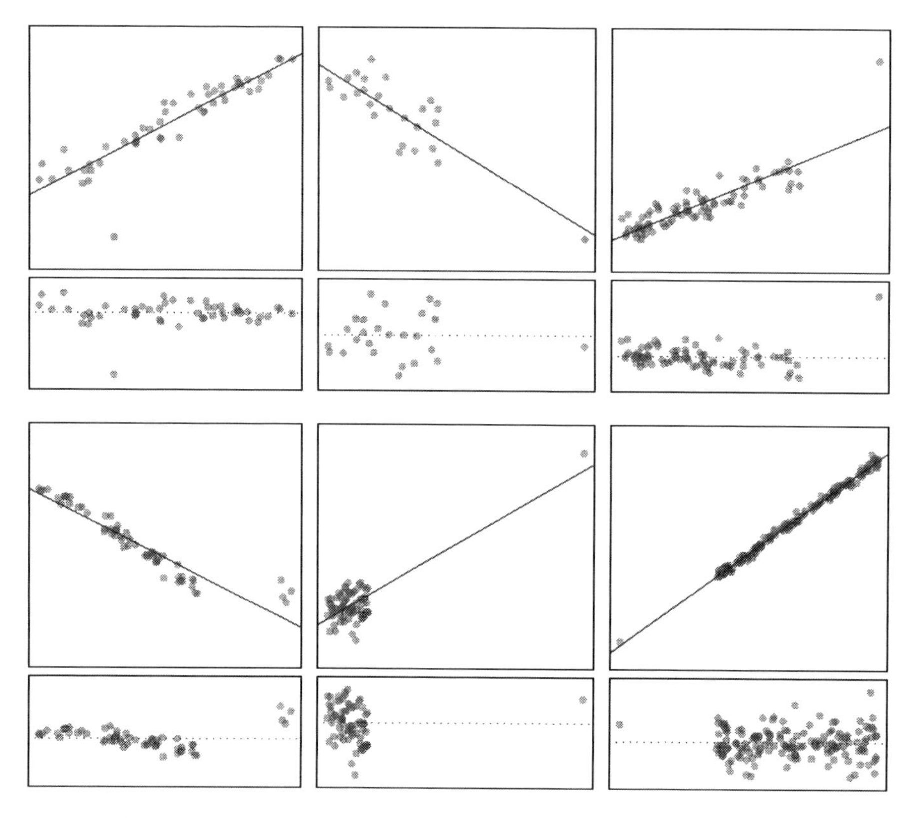

Figure 7.16: Six plots, each with a least squares line and residual plot. All data sets have at least one outlier.

Examine the residual plots in Figure 7.16. You will probably find that there is some trend in the main clouds of (3) and (4). In these cases, the outlier was influential on the slope of the least squares line. In (5), data with no clear trend was assigned a line with a large trend simply due to one outlier (!).

Leverage
Points that fall, horizontally, away from the center of the cloud tend to pull harder on the line, so we call them points with **high leverage**.

Points that fall horizontally far from the line are points of high leverage; these points can strongly influence the slope of the least squares line. If one of these high leverage points does appear to actually invoke its influence on the slope of the line – as in cases (3), (4), and (5) of Exercise 7.11 – then we call it an **influential point**. Usually we can say a point is influential if, had we fit the line without it, the influential point would have been unusually far from the least squares line.

It is tempting to remove outliers. Don't do this without very good reason. Models that ignore exceptional (and interesting) cases often perform poorly. For instance, if a financial firm ignored the largest market swings – the "outliers" – they would soon go bankrupt by making poorly thought-out investments.

Caution: Don't ignore outliers when fitting a final model
If there are outliers in the data, they should not be removed or ignored without good reason. Whatever final model is fit to the data would not be very helpful if it ignores the most exceptional cases.

7.4 Inference for linear regression

Authors' note: This section will be completed at a later date.

then a small secondary cloud of four outliers. The secondary cloud appears to be influencing the line somewhat strongly, making the least square line fit poorly almost everywhere. There might be an interesting explanation for the dual clouds, which is something that could be investigated. (5) there is no obvious trend in the main cloud of points and the outlier on the right appears to largely control the slope of the least squares line. (6) There is one outlier far from the cloud, however, it falls quite close to the least squares line and does not appear to be very influential.

7.5 Problem set

7.5.1 Line fitting, residuals, and correlation

7.1 For each scatterplot given below with a regression superimposed, describe how the residuals plot (residuals versus x) would look like.

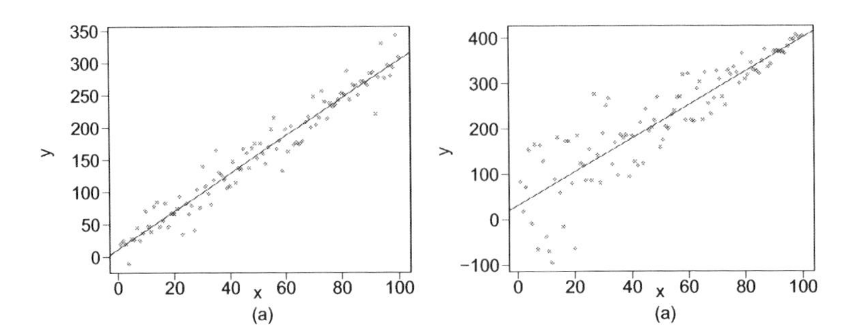

7.2 For the residual plots given below, describe any apparent trends and determine if a linear model would be appropriate for these data. Explain your reasoning.

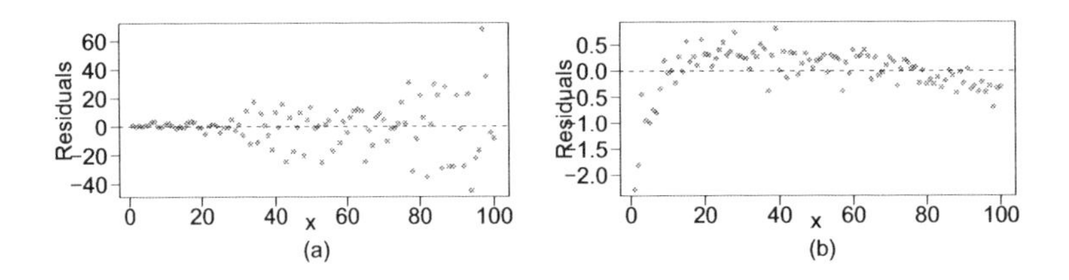

7.3 Describe the strength of the relationship of the pair of variables in each plot (e.g. weak, moderate, or strong). Which of these are linear relationships where we might try fitting a linear regression line?

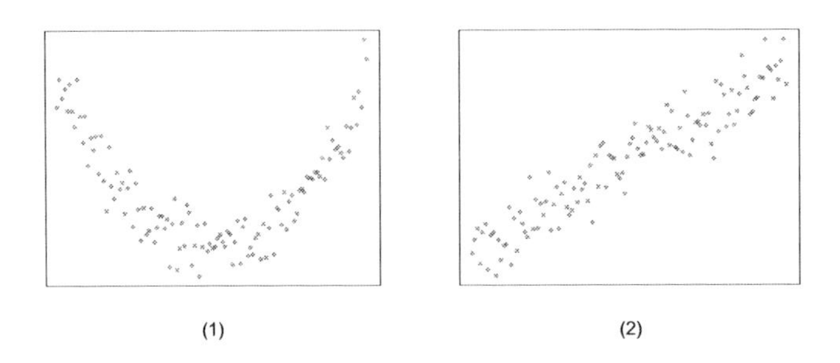

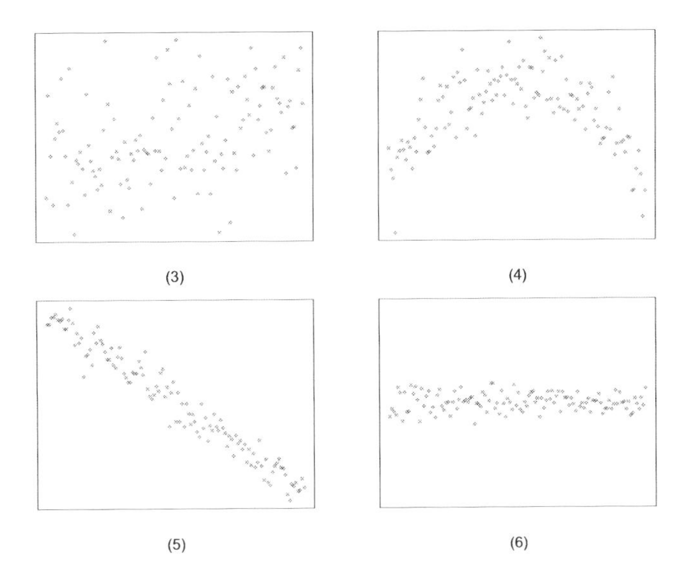

7.4 Describe the strength of the relationship of the pair of variables in each plot (e.g. weak, moderate, or strong). Which of these are linear relationships where we might try fitting a linear regression line?

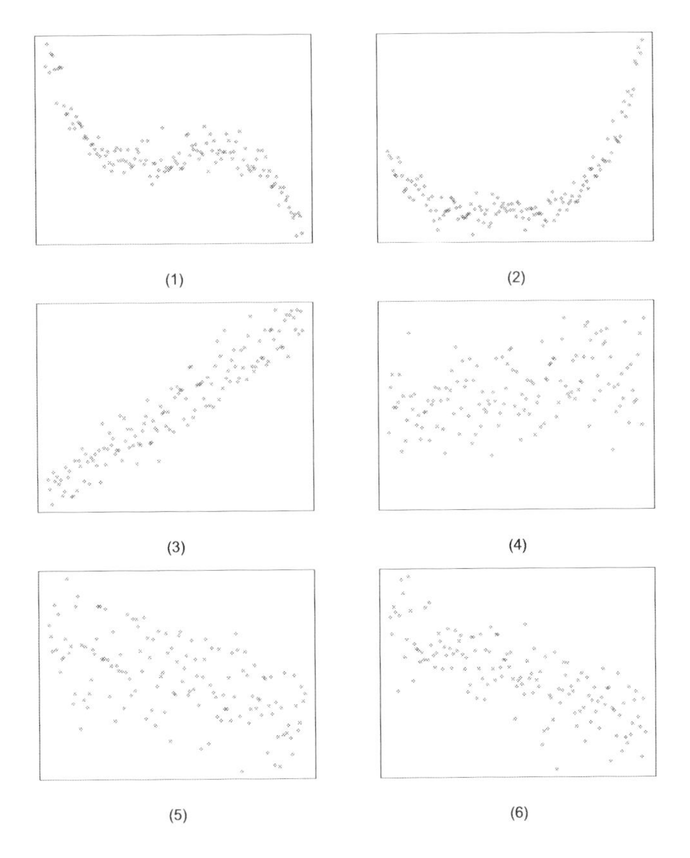

7.5 Match the calculated correlations to the corresponding scatterplot.

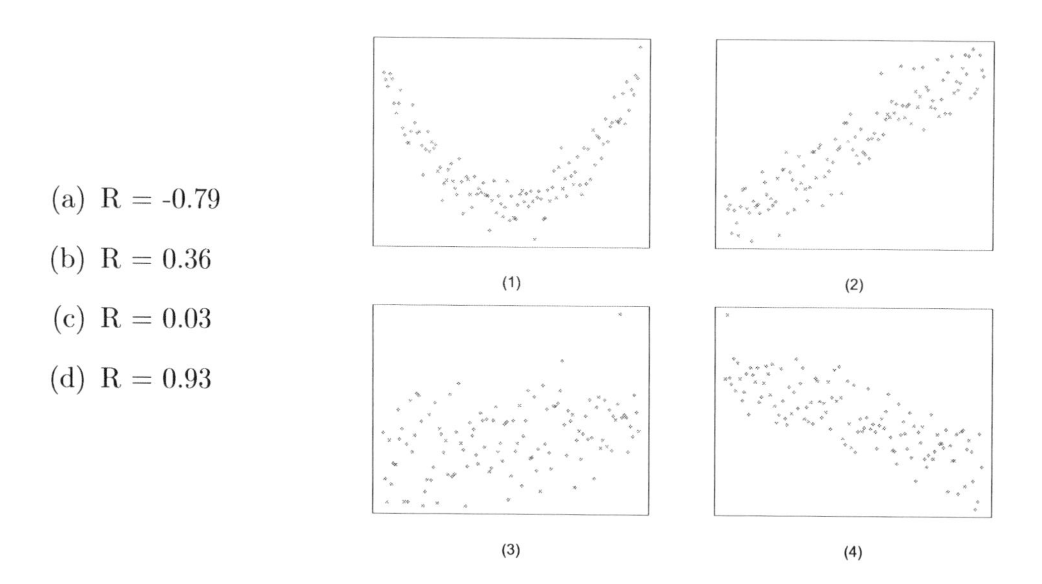

(a) R = -0.79

(b) R = 0.36

(c) R = 0.03

(d) R = 0.93

7.6 Match the calculated correlations to the corresponding scatterplot.

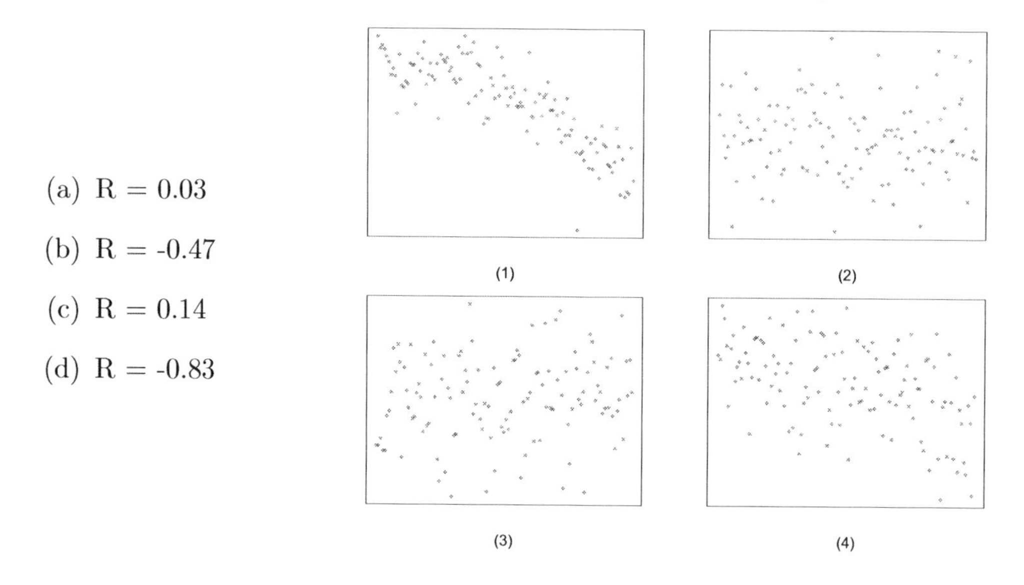

(a) R = 0.03

(b) R = -0.47

(c) R = 0.14

(d) R = -0.83

7.7 Below are two scatterplots based on grades recorded during several years for a Statistics course at a university. The first scatterplot shows the association between final exam grade and first exam grade for 233 students, and the second scatterplot shows the association between final exam grade and second exam grade.

(a) Based on these graphs which of the two exams has the strongest correlation with the final exam grade? Explain.

(b) Can you think of a reason why the correlation between the exam you chose in part (a) and the final exam is higher?

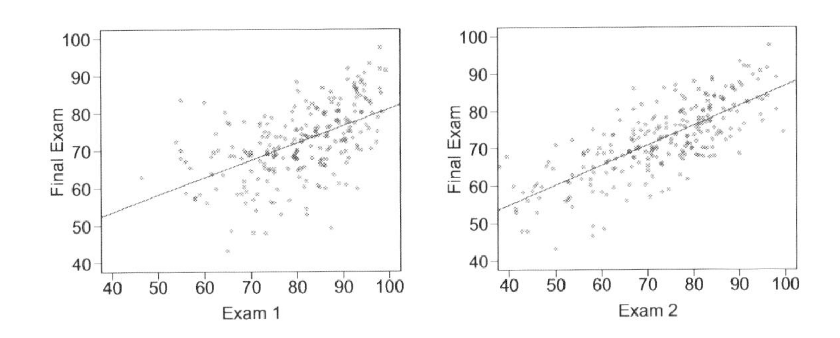

7.8 The Coast Starlight Amtrak train runs from Seattle to Los Angeles. The scatterplot below displays the distance between each stop (in miles) and the amount of time it takes to travel from one stop to another (in minutes).

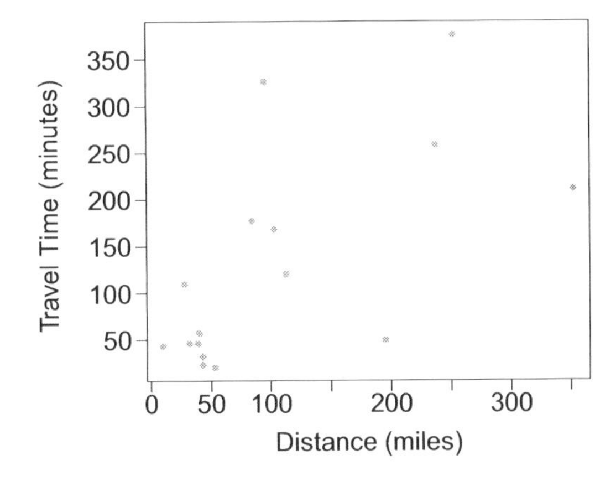

(a) Describe the relationship between distance and travel time.

(b) How would the relationship change if travel time was instead measured in hours and distance was instead measured in kilometers.

(c) Correlation between travel time (in miles) and distance (in minutes), R = 0.636. What is the correlation between travel time (in minutes) and distance (in hours).

7.9 What would be the correlation between the ages of husband and wife if men always married woman who were

(a) 3 years younger than themselves?

(b) 2 years older than themselves?

(c) half as old as themselves?

7.5.2 Fitting a line by least squares regression

7.10 Association of Turkish Travel Agencies reports the number of foreign tourists visiting Turkey and tourist spending by year [?]. The below scatterplot shows the relationship between these two variables along with the least squares fit.

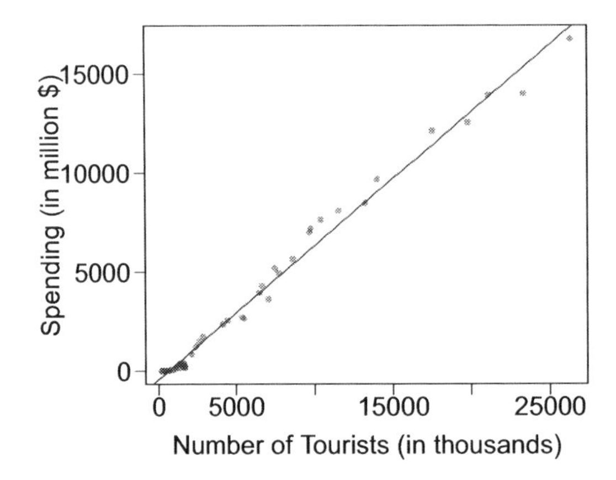

(a) Describe the relationship between number of tourists and spending.

(b) What are the explanatory and response variables?

(c) Why might we want to fit a regression line to this data?

7.11 Does the tourism data plotted in Exercise 10 meet the conditions required for fitting a least squares line? In addition to the scatterplot provided in Exercise 10, use the residuals plot and the histogram below to answer this question.

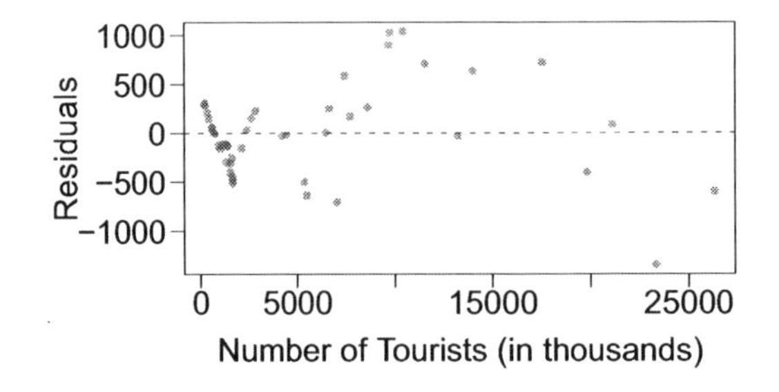

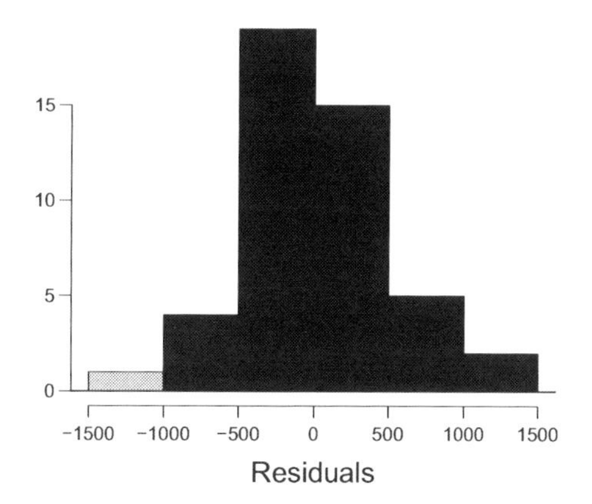

7.12 Exercise 8 introduces data on the Coast Starlight Amtrak train that runs from Seattle to Los Angeles. The mean travel time from one stop to the next on the Coast Starlight is 129 mins, with a standard deviation of 113 minutes. The mean distance travelled from one stop to the next is 107 miles with a standard deviation of 99 miles. The correlation between travel time and distance is 0.636.

(a) Write the equation of the regression line for predicting travel time from distance travelled.

(b) Interpret the slope and the intercept in context.

7.13 The distance between Santa Barbara and Los Angeles is 103 miles. Calculate the following based on the linear model from Exercise 12.

(a) How long does the linear model predict that it would take the Coast Starlight to travel this distance?

(b) It actually takes the the Coast Starlight 168 mins to travel from Santa Barbara to Los Angeles. Calculate the residual and explain what a negative/positive residual means.

7.14 Amtrak is considering adding a stop to the Coast Starlight 500 miles away from Los Angeles. Use the model from Exercise 12 to predict the travel time from Los Angeles to this point.

7.15 Based on the information given in Exercise 12, calculate R^2 of the regression line for predicting travel dime from distance travelled for the Coast Starlight and interpret it in context.

7.5.3 Types of outliers in linear regression

7.16 Identify the outliers in the scatter plots shown below and determine what type of an outlier it is? Explain your reasoning.

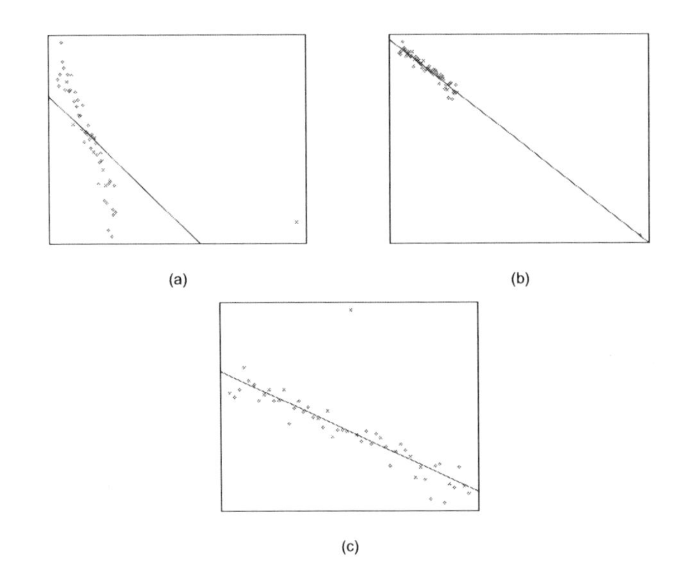
(a)
(b)
(c)

Appendix A

Distribution tables

A.1 Normal Probability Table

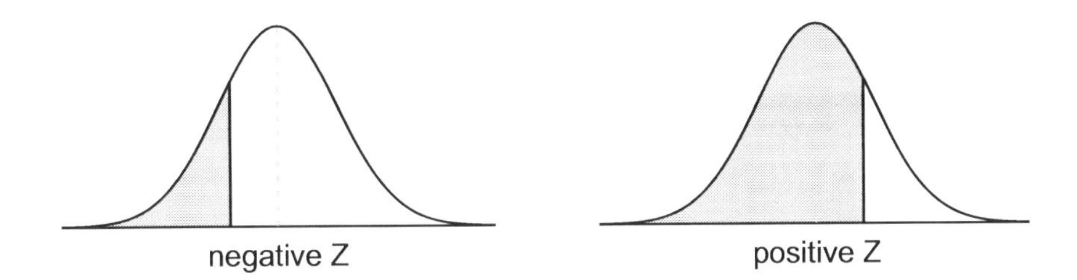

Figure A.1: The area to the left of Z represents the percentile of the observation. The normal probability table always lists percentiles.

Second decimal place of Z										
0.09	0.08	0.07	0.06	0.05	0.04	0.03	0.02	0.01	0.00	Z
0.0002	0.0003	0.0003	0.0003	0.0003	0.0003	0.0003	0.0003	0.0003	0.0003	−3.4
0.0003	0.0004	0.0004	0.0004	0.0004	0.0004	0.0004	0.0005	0.0005	0.0005	−3.3
0.0005	0.0005	0.0005	0.0006	0.0006	0.0006	0.0006	0.0006	0.0007	0.0007	−3.2
0.0007	0.0007	0.0008	0.0008	0.0008	0.0008	0.0009	0.0009	0.0009	0.0010	−3.1
0.0010	0.0010	0.0011	0.0011	0.0011	0.0012	0.0012	0.0013	0.0013	0.0013	−3.0
0.0014	0.0014	0.0015	0.0015	0.0016	0.0016	0.0017	0.0018	0.0018	0.0019	−2.9
0.0019	0.0020	0.0021	0.0021	0.0022	0.0023	0.0023	0.0024	0.0025	0.0026	−2.8
0.0026	0.0027	0.0028	0.0029	0.0030	0.0031	0.0032	0.0033	0.0034	0.0035	−2.7
0.0036	0.0037	0.0038	0.0039	0.0040	0.0041	0.0043	0.0044	0.0045	0.0047	−2.6
0.0048	0.0049	0.0051	0.0052	0.0054	0.0055	0.0057	0.0059	0.0060	0.0062	−2.5
0.0064	0.0066	0.0068	0.0069	0.0071	0.0073	0.0075	0.0078	0.0080	0.0082	−2.4
0.0084	0.0087	0.0089	0.0091	0.0094	0.0096	0.0099	0.0102	0.0104	0.0107	−2.3
0.0110	0.0113	0.0116	0.0119	0.0122	0.0125	0.0129	0.0132	0.0136	0.0139	−2.2
0.0143	0.0146	0.0150	0.0154	0.0158	0.0162	0.0166	0.0170	0.0174	0.0179	−2.1
0.0183	0.0188	0.0192	0.0197	0.0202	0.0207	0.0212	0.0217	0.0222	0.0228	−2.0
0.0233	0.0239	0.0244	0.0250	0.0256	0.0262	0.0268	0.0274	0.0281	0.0287	−1.9
0.0294	0.0301	0.0307	0.0314	0.0322	0.0329	0.0336	0.0344	0.0351	0.0359	−1.8
0.0367	0.0375	0.0384	0.0392	0.0401	0.0409	0.0418	0.0427	0.0436	0.0446	−1.7
0.0455	0.0465	0.0475	0.0485	0.0495	0.0505	0.0516	0.0526	0.0537	0.0548	−1.6
0.0559	0.0571	0.0582	0.0594	0.0606	0.0618	0.0630	0.0643	0.0655	0.0668	−1.5
0.0681	0.0694	0.0708	0.0721	0.0735	0.0749	0.0764	0.0778	0.0793	0.0808	−1.4
0.0823	0.0838	0.0853	0.0869	0.0885	0.0901	0.0918	0.0934	0.0951	0.0968	−1.3
0.0985	0.1003	0.1020	0.1038	0.1056	0.1075	0.1093	0.1112	0.1131	0.1151	−1.2
0.1170	0.1190	0.1210	0.1230	0.1251	0.1271	0.1292	0.1314	0.1335	0.1357	−1.1
0.1379	0.1401	0.1423	0.1446	0.1469	0.1492	0.1515	0.1539	0.1562	0.1587	−1.0
0.1611	0.1635	0.1660	0.1685	0.1711	0.1736	0.1762	0.1788	0.1814	0.1841	−0.9
0.1867	0.1894	0.1922	0.1949	0.1977	0.2005	0.2033	0.2061	0.2090	0.2119	−0.8
0.2148	0.2177	0.2206	0.2236	0.2266	0.2296	0.2327	0.2358	0.2389	0.2420	−0.7
0.2451	0.2483	0.2514	0.2546	0.2578	0.2611	0.2643	0.2676	0.2709	0.2743	−0.6
0.2776	0.2810	0.2843	0.2877	0.2912	0.2946	0.2981	0.3015	0.3050	0.3085	−0.5
0.3121	0.3156	0.3192	0.3228	0.3264	0.3300	0.3336	0.3372	0.3409	0.3446	−0.4
0.3483	0.3520	0.3557	0.3594	0.3632	0.3669	0.3707	0.3745	0.3783	0.3821	−0.3
0.3859	0.3897	0.3936	0.3974	0.4013	0.4052	0.4090	0.4129	0.4168	0.4207	−0.2
0.4247	0.4286	0.4325	0.4364	0.4404	0.4443	0.4483	0.4522	0.4562	0.4602	−0.1
0.4641	0.4681	0.4721	0.4761	0.4801	0.4840	0.4880	0.4920	0.4960	0.5000	−0.0

*For $Z \leq -3.50$, the probability is less than or equal to 0.0002.

	Second decimal place of Z									
Z	0.00	0.01	0.02	0.03	0.04	0.05	0.06	0.07	0.08	0.09
0.0	0.5000	0.5040	0.5080	0.5120	0.5160	0.5199	0.5239	0.5279	0.5319	0.5359
0.1	0.5398	0.5438	0.5478	0.5517	0.5557	0.5596	0.5636	0.5675	0.5714	0.5753
0.2	0.5793	0.5832	0.5871	0.5910	0.5948	0.5987	0.6026	0.6064	0.6103	0.6141
0.3	0.6179	0.6217	0.6255	0.6293	0.6331	0.6368	0.6406	0.6443	0.6480	0.6517
0.4	0.6554	0.6591	0.6628	0.6664	0.6700	0.6736	0.6772	0.6808	0.6844	0.6879
0.5	0.6915	0.6950	0.6985	0.7019	0.7054	0.7088	0.7123	0.7157	0.7190	0.7224
0.6	0.7257	0.7291	0.7324	0.7357	0.7389	0.7422	0.7454	0.7486	0.7517	0.7549
0.7	0.7580	0.7611	0.7642	0.7673	0.7704	0.7734	0.7764	0.7794	0.7823	0.7852
0.8	0.7881	0.7910	0.7939	0.7967	0.7995	0.8023	0.8051	0.8078	0.8106	0.8133
0.9	0.8159	0.8186	0.8212	0.8238	0.8264	0.8289	0.8315	0.8340	0.8365	0.8389
1.0	0.8413	0.8438	0.8461	0.8485	0.8508	0.8531	0.8554	0.8577	0.8599	0.8621
1.1	0.8643	0.8665	0.8686	0.8708	0.8729	0.8749	0.8770	0.8790	0.8810	0.8830
1.2	0.8849	0.8869	0.8888	0.8907	0.8925	0.8944	0.8962	0.8980	0.8997	0.9015
1.3	0.9032	0.9049	0.9066	0.9082	0.9099	0.9115	0.9131	0.9147	0.9162	0.9177
1.4	0.9192	0.9207	0.9222	0.9236	0.9251	0.9265	0.9279	0.9292	0.9306	0.9319
1.5	0.9332	0.9345	0.9357	0.9370	0.9382	0.9394	0.9406	0.9418	0.9429	0.9441
1.6	0.9452	0.9463	0.9474	0.9484	0.9495	0.9505	0.9515	0.9525	0.9535	0.9545
1.7	0.9554	0.9564	0.9573	0.9582	0.9591	0.9599	0.9608	0.9616	0.9625	0.9633
1.8	0.9641	0.9649	0.9656	0.9664	0.9671	0.9678	0.9686	0.9693	0.9699	0.9706
1.9	0.9713	0.9719	0.9726	0.9732	0.9738	0.9744	0.9750	0.9756	0.9761	0.9767
2.0	0.9772	0.9778	0.9783	0.9788	0.9793	0.9798	0.9803	0.9808	0.9812	0.9817
2.1	0.9821	0.9826	0.9830	0.9834	0.9838	0.9842	0.9846	0.9850	0.9854	0.9857
2.2	0.9861	0.9864	0.9868	0.9871	0.9875	0.9878	0.9881	0.9884	0.9887	0.9890
2.3	0.9893	0.9896	0.9898	0.9901	0.9904	0.9906	0.9909	0.9911	0.9913	0.9916
2.4	0.9918	0.9920	0.9922	0.9925	0.9927	0.9929	0.9931	0.9932	0.9934	0.9936
2.5	0.9938	0.9940	0.9941	0.9943	0.9945	0.9946	0.9948	0.9949	0.9951	0.9952
2.6	0.9953	0.9955	0.9956	0.9957	0.9959	0.9960	0.9961	0.9962	0.9963	0.9964
2.7	0.9965	0.9966	0.9967	0.9968	0.9969	0.9970	0.9971	0.9972	0.9973	0.9974
2.8	0.9974	0.9975	0.9976	0.9977	0.9977	0.9978	0.9979	0.9979	0.9980	0.9981
2.9	0.9981	0.9982	0.9982	0.9983	0.9984	0.9984	0.9985	0.9985	0.9986	0.9986
3.0	0.9987	0.9987	0.9987	0.9988	0.9988	0.9989	0.9989	0.9989	0.9990	0.9990
3.1	0.9990	0.9991	0.9991	0.9991	0.9992	0.9992	0.9992	0.9992	0.9993	0.9993
3.2	0.9993	0.9993	0.9994	0.9994	0.9994	0.9994	0.9994	0.9995	0.9995	0.9995
3.3	0.9995	0.9995	0.9995	0.9996	0.9996	0.9996	0.9996	0.9996	0.9996	0.9997
3.4	0.9997	0.9997	0.9997	0.9997	0.9997	0.9997	0.9997	0.9997	0.9997	0.9998

*For $Z \geq 3.50$, the probability is greater than or equal to 0.9998.

A.2 t Distribution Table

one tail	0.100	0.050	0.025	0.010	0.005
two tails	0.200	0.100	0.050	0.020	0.010
df 1	3.08	6.31	12.71	31.82	63.66
2	1.89	2.92	4.30	6.96	9.92
3	1.64	2.35	3.18	4.54	5.84
4	1.53	2.13	2.78	3.75	4.60
5	1.48	2.02	2.57	3.36	4.03
6	1.44	1.94	2.45	3.14	3.71
7	1.41	1.89	2.36	3.00	3.50
8	1.40	1.86	2.31	2.90	3.36
9	1.38	1.83	2.26	2.82	3.25
10	1.37	1.81	2.23	2.76	3.17
11	1.36	1.80	2.20	2.72	3.11
12	1.36	1.78	2.18	2.68	3.05
13	1.35	1.77	2.16	2.65	3.01
14	1.35	1.76	2.14	2.62	2.98
15	1.34	1.75	2.13	2.60	2.95
16	1.34	1.75	2.12	2.58	2.92
17	1.33	1.74	2.11	2.57	2.90
18	1.33	1.73	2.10	2.55	2.88
19	1.33	1.73	2.09	2.54	2.86
20	1.33	1.72	2.09	2.53	2.85
21	1.32	1.72	2.08	2.52	2.83
22	1.32	1.72	2.07	2.51	2.82
23	1.32	1.71	2.07	2.50	2.81
24	1.32	1.71	2.06	2.49	2.80
25	1.32	1.71	2.06	2.49	2.79
26	1.31	1.71	2.06	2.48	2.78
27	1.31	1.70	2.05	2.47	2.77
28	1.31	1.70	2.05	2.47	2.76
29	1.31	1.70	2.05	2.46	2.76
30	1.31	1.70	2.04	2.46	2.75

one tail	0.100	0.050	0.025	0.010	0.005
two tails	0.200	0.100	0.050	0.020	0.010
df 31	1.31	1.70	2.04	2.45	2.74
32	1.31	1.69	2.04	2.45	2.74
33	1.31	1.69	2.03	2.44	2.73
34	1.31	1.69	2.03	2.44	2.73
35	1.31	1.69	2.03	2.44	2.72
36	1.31	1.69	2.03	2.43	2.72
37	1.30	1.69	2.03	2.43	2.72
38	1.30	1.69	2.02	2.43	2.71
39	1.30	1.68	2.02	2.43	2.71
40	1.30	1.68	2.02	2.42	2.70
41	1.30	1.68	2.02	2.42	2.70
42	1.30	1.68	2.02	2.42	2.70
43	1.30	1.68	2.02	2.42	2.70
44	1.30	1.68	2.02	2.41	2.69
45	1.30	1.68	2.01	2.41	2.69
46	1.30	1.68	2.01	2.41	2.69
47	1.30	1.68	2.01	2.41	2.68
48	1.30	1.68	2.01	2.41	2.68
49	1.30	1.68	2.01	2.40	2.68
50	1.30	1.68	2.01	2.40	2.68
55	1.30	1.67	2.00	2.40	2.67
60	1.30	1.67	2.00	2.39	2.66
65	1.29	1.67	2.00	2.39	2.65
70	1.29	1.67	1.99	2.38	2.65
75	1.29	1.67	1.99	2.38	2.64
80	1.29	1.66	1.99	2.37	2.64
85	1.29	1.66	1.99	2.37	2.63
90	1.29	1.66	1.99	2.37	2.63
95	1.29	1.66	1.99	2.37	2.63
100	1.29	1.66	1.98	2.36	2.63
120	1.29	1.66	1.98	2.36	2.62
140	1.29	1.66	1.98	2.35	2.61
160	1.29	1.65	1.97	2.35	2.61
180	1.29	1.65	1.97	2.35	2.60
200	1.29	1.65	1.97	2.35	2.60
300	1.28	1.65	1.97	2.34	2.59
400	1.28	1.65	1.97	2.34	2.59
500	1.28	1.65	1.96	2.33	2.59
∞	1.28	1.64	1.96	2.33	2.58

Index

Bibliography

[1] B. Ritz, F. Yu, G. Chapa, and S. Fruin, "Effect of air pollution on preterm birth among children born in Southern California between 1989 and 1993," *Epidemiology*, vol. 11, no. 5, pp. 502–511, 2000.

[2] J. McGowan, "Health Education: Does the Buteyko Institute Method make a difference?," *Thorax*, vol. 58, 2003.

[3] Source: www.stats4schools.gov.uk, November 10, 2009.

[4] R. Fisher, "The use of multiple measurements in taxonomic problems," *Annals of Eugenics*, vol. 7, pp. 179–188, 1936.

[5] T. Allison and D. Cicchetti, "Sleep in mammals: ecological and constitutional correlates," *Arch. Hydrobiol*, vol. 75, p. 442, 1975.

[6] B. Turnbull, B. Brown, and M. Hu, "Survivorship of heart transplant data," *Journal of the American Statistical Association*, vol. 69, pp. 74–80, 1974.

[7] Source: http://babymed.com/Tools/Other/eye_color/Default.aspx.

[8] Source: SAMHSA, Office of Applied Studies, National Survey on Drug Use and Health, 2007 and 2008, http://www.oas.samhsa.gov/NSDUH/2k8NSDUH/tabs/Sect2peTabs1to42.htm#Tab2.5B.

[9] Source: http://www.ets.org/Media/Tests/GRE/pdf/gre_0809_interpretingscores.pdf.

[10] A. Romero-Corral, V. Somers, J. Sierra-Johnson, R. Thomas, M. Collazo-Clavell, J. Korinek, T. Allison, J. Batsis, F. Sert-Kuniyoshi, and F. Lopez-Jimenez, "Accuracy of body mass index in diagnosing obesity in the adult general population," *International Journal of Obesity*, vol. 32, no. 6, pp. 959–966, 2008.

[11] Public option gains support, October 20, 2009.

[12] Perceived Insufficient Rest or Sleep Among Adults United States, 2008, October 30, 2009.

[13] Source: http://stat.ethz.ch/R-manual/R-patched/library/datasets/html/chickwts.html.

[14] Source: http://www.fueleconomy.gov/feg/download.shtml.

Made in the USA
Lexington, KY
31 January 2011